SUSTAINING
MARINE FISHERIES

Committee on Ecosystem Management for Sustainable Marine Fisheries
Ocean Studies Board
Commission on Geosciences, Environment, and Resources
National Research Council

NATIONAL ACADEMY PRESS
Washington D.C.   1999

NATIONAL ACADEMY PRESS • 2101 Constitution Avenue, NW • Washington, DC 20418

NOTICE: The project that is the subject of this report was approved by the Governing Board of the National Research Council, whose members are drawn from the councils of the National Academy of Sciences, the National Academy of Engineering, and the Institute of Medicine. The members of the committee responsible for the report were chosen for their special competency and with regard for appropriate balance.

This study was a Governing Board Theme Initiative project and funded by the Academy Industry Program Fund, the Mellon Fund, and the Casey Fund of the National Research Council, and the Kellogg Endowment Fund of the National Academy of Sciences and the Institute of Medicine. Any opinions, findings, conclusions, or recommendations expressed in this publication are those of the authors and do not necessarily reflect the view of the organizations or agencies that provided support for the project.

Cover art was created by Alfredo M. Arreguin. Mr. Arreguin is an internationally recognized artist who lives in Seattle, Washington. For many years he has painted the world's endangered ecosystems—the jungles and wetlands, as well as the salmon of the Pacific Northwest. His work is displayed in numerous collections, including the National Museum of American Art in Washington, D.C.

Library of Congress Cataloging-in-Publication Data

Sustaining marine fisheries / Committee on Ecosystem Management for
Sustainable Marine Fisheries, Ocean Studies Board, Commission on
Geosciences, Environment, and Resources, National Research Council.
    p. cm.
    Includes bibliographical references and index.
    ISBN 0-309-05526-1 (casebound)
    1. Sustainable fisheries. 2. Fishery management. 3. Marine
ecology. I. National Research Council (U.S.). Committee on
Ecosystem Management for Sustainable Marine Fisheries.
    SH329.S87 S87 1998
    639.3'2—dc21
                                  98-58059

Sustaining Marine Fisheries is available from the National Academy Press, 2101 Constitution Avenue, N.W., Lockbox 285, Washington, DC 20055; 800-624-6242 or 202-334-3313 (in the Washington metropolitan area); http://www.nap.edu

Printed in the United States of America

This report is dedicated
to the memory of committee member
Nathaniel Bingham (1938-1998)

The National Academy of Sciences is a private, nonprofit, self-perpetuating society of distinguished scholars engaged in scientific and engineering research, dedicated to the furtherance of science and technology and to their use for the general welfare. Upon the authority of the charter granted to it by the Congress in 1863, the Academy has a mandate that requires it to advise the federal government on scientific and technical matters. Dr. Bruce M. Alberts is president of the National Academy of Sciences.

The National Academy of Engineering was established in 1964, under the charter of the National Academy of Sciences, as a parallel organization of outstanding engineers. It is autonomous in its administration and in the selection of its members, sharing with the National Academy of Sciences the responsibility of advising the federal government. The National Academy of Engineering also sponsors engineering programs aimed at meeting national needs, encourages education and research, and recognizes the superior achievements of engineers. Dr. William A. Wulf is president of the National Academy of Engineering.

The Institute of Medicine was established in 1970 by the National Academy of Sciences to secure the services of eminent members of appropriate professions in the examination of policy matters pertaining to the health of the public. The Institute acts under the responsibility given to the National Academy of Sciences by its congressional charter to be an adviser to the federal government and, upon its own initiative, to identify issues of medical care, research, and education. Dr. Kenneth I. Shine is president of the Institute of Medicine.

The National Research Council was organized by the National Academy of Sciences in 1916 to associate the broad community of science and technology with the Academy's purposes of furthering knowledge and advising the federal government. Functioning in accordance with general policies determined by the Academy, the Council has become the principal operating agency of both the National Academy of Sciences and the National Academy of Engineering in providing services to the government, the public, and the scientific and engineering communities. The Council is administered jointly by both Academies and the Institute of Medicine. Dr. Bruce M. Alberts and Dr. William A. Wulf are chairman and vice chairman, respectively, of the National Research Council.

OCEAN STUDIES BOARD

Acknowledgment of Reviewers

This report has been reviewed in draft form by individuals chosen for their diverse perspectives and technical expertise, in accordance with procedures approved by the NRC's Report Review Committee. The purpose of this independent review is to provide candid and critical comments that will assist the NRC in making the published report as sound as possible and to ensure that the report meets institutional standards for objectivity, evidence, and responsiveness to the study charge. The content of the final report is the responsibility of the NRC and the study committee, and not the responsibility of the reviewers. The review comments and draft manuscript remain confidential to protect the integrity of the deliberative process. We wish to thank the following individuals for their participation in the review of this report:

Chris Blackburn, Alaska Groundfish Data Bank
John Chipman, University of Minnesota
Ellie Dorsey, Conservation Law Foundation
Richard Haedrich, Memorial University of Newfoundland
Susan Hanna, Oregon State University
Raymond Hilborn, University of Washington
James Kitchell, University of Wisconsin, Madison
John Ledyard, California Institute of Technology
Kai Lee, Williams College
Pamela Matson, Stanford University
Ransom Myers, Dalhousie University

William Pearcy, Oregon State University
C.H. Peterson, University of North Carolina, Chapel Hill
Terrance Quinn II, University of Alaska Fairbanks, Juneau Center

While the individuals listed above have provided many constructive comments and suggestions, it must be emphasized that responsibility for the
final content of this report rests entirely with the authoring committee and
the NRC.

Foreword

Fishery issues continue to receive enormous, and growing, public attention. A particularly good example can be seen in the groundfish fisheries off New England, where increasingly stringent regulations have been implemented to limit the capture of cod, haddock, flounder, and other fishes. Many other marine fisheries are similarly troubled. Yet, despite considerable study, the exact causes of the problems and the means to solve them are often difficult to understand.

The National Research Council's Ocean Studies Board (OSB) has been actively involved in a number of studies related to marine fisheries, leading to such reports as An Assessment of Atlantic Bluefin Tuna (1994), Improving the Management of U.S. Marine Fisheries (1994), and Improving Fish Stock Assessments (1998). The issues presented by studies such as these highlight the need for taking a broad view of fishery problems. Thus, the Ocean Studies Board designed the study that is the subject of this report, assembling a group of experts to produce the broad-based overview presented here. In addition, several topics raised in this volume are currently being explored in greater detail by ongoing OSB study committees including the Committee to Review Individual Fishing Quotas, the Committee to Review Community Development Quotas, the Committee on the Evaluation, Design, and Monitoring of Marine Reserves and Protected Areas in the U.S., and the Committee on Improving the Collection and Use of Fisheries Data.

We look forward to continuing to make the connections between fishery science and policy that are necessary to achieve sustainable resource management.

Kenneth Brink, *Chairman*
Ocean Studies Board

Preface

Producing this report was a difficult challenge because of the complexity of the issue—that of trying to bring new insights and approaches into the ways that fisheries are viewed and managed. The need for this evaluation is clear. Many of the fisheries of the world's oceans are under threat. These threatened fisheries are important economically, culturally, and for supplying protein to a growing human population. The ecosystems to which the targeted fish, invertebrates, and plants belong provide additional goods and services to society, so they too must be considered in a holistic view of the problem.

It is this holistic viewpoint that we have sought. The Ocean Studies Board committee that produced this report was unusually broad in its expertise and included fishery scientists, ecosystem and population ecologists, fishers, and social scientists, including economists. Its membership includes people from the fishing industry and from nongovernmental organizations. As can be imagined, achieving a convergence of viewpoints among such a diverse group was challenging. However, it is just such a convergence that is necessary, as discussed in this report, to open new approaches to the difficult problem of sustaining marine fisheries.

In addition to the direct input of committee members, we sought advice at a conference in Monterey, California, from a larger group of international experts representing, again, a diversity of approaches. The results of the discussions at that meeting were presented in a recent special issue of *Ecological Applications* and also importantly influenced the committee's deliberations, as reflected in this report.

The committee also sought advice and insight from a group of fishers and conservationists at two discussion groups, one in Seattle and the other in Washington, D.C. These meetings clearly indicated the history of the problems and the universal desire to find equitable and achievable ways of addressing the issue of the health of fishery resources.

As will be seen in this report, the committee has no silver bullet to offer. The problem is too large and too complex for a single solution. What we do offer is an overview of the problem and the history of its development. We do point to some pervasive parts of the problem that must be addressed and then offer specific approaches, many of which are already in place, that need amplification and further development. Most of all, the committee proposes a new context, an ecosystem viewpoint in which humans are the major player, in which we must proceed in order to have any hope of maintaining sustainable fisheries in a world in which climate is changing and the human population is growing.

Many individuals and organizations helped the committee in its work. We are grateful to scientists around the world who provided us with information, literature citations, and advice. The National Marine Fisheries Service and the United Nations' Food and Agriculture Organization were particularly helpful with documentation.

This study was stimulated by the actions of Mary Hope Katsouros, former director of the Ocean Studies Board. For her extraordinary energy and insights the committee is grateful. We are pleased that the National Research Council (NRC) agreed that this problem is so important that it funded this study from internal sources. The committee also thanks NRC Chair Bruce Alberts and NRC Executive Officer E. William Colglazier for their personal interest in and help with this project. The staff of the Ocean Studies Board provided the usual excellent backup for the project. The committee is especially grateful to project officer David Policansky, the quintessential professional, for his never-flagging, crucial, and substantial input in bringing this report to fruition.

Harold A. Mooney, *Chairman*
Committee on Ecosystem Management for
Sustainable Marine Fisheries

Contents

Executive Summary

Marine ecosystems are being perturbed by fishing and other human activities. Many marine fisheries are in decline, and the effects of fishing on other ecosystem goods and services[1] are beginning to be understood and recognized. In recent years, global marine catches appear to have reached a plateau of about 84 million metric tons[2] per year, although total fish production, which includes aquaculture, has continued to increase. In some cases, fisheries have been entirely closed, and in many others it takes increasing effort to maintain catch rates. Fishing is also an economically important international industry, with first-sale revenues of approximately $U.S. 100 billion per year for all fishery products. (Farm-raised and freshwater fisheries account for approximately 25 percent by weight of all fishery products.) Globally, fishery products directly provided approximately 14 kg of food per person in 1996; approximately 28 percent of global fishery products was used for animal feed and other products that do not contribute directly to human food. Although in recent years total fish production has increased faster than the human population, the total from marine-capture fisheries has increased little if at all.

To evaluate whether current marine-capture fisheries are sustainable, to determine to what degree marine ecosystems are affected by fishing, and to assess whether an ecosystem approach to fishery management can help achieve

---

[1]Ecosystem goods and services are those ecosystem products and processes that directly benefit humans. They include food, breathable air, clean water, fiber, medicines, quality of life, and many other items.

[2]One metric ton, usually abbreviated t, is 1,000 kg, approximately 2,205 lbs.

sustainability, the National Research Council's Ocean Studies Board established the Committee on Ecosystem Management for Sustainable Marine Fisheries. The committee was directed to "assess the current state of fisheries resources; the basis for success and failure in marine fisheries management (including the role of science); and the implications of fishery activities to ecosystem structure and function. Each activity [was to] be considered relative to sustaining populations of fish and other marine resources" (Statement of Task). This report is the product of the committee's study.

SUSTAINABILITY AND ECOSYSTEM-BASED MANAGEMENT

The sea was long viewed as an inexhaustible supply of protein for human use. But recently, as the potential and actual adverse effects of human activities have become apparent, our views of marine ecosystems have changed. It has become increasingly clear that the ocean's resources are not inexhaustible. And, in addition to direct societal benefits from fishing, ecosystem goods and services have become recognized as valuable and irreplaceable natural resources. These insights have led to a concern regarding sustainability and an interest in the potential of ecosystem-based approaches to fishery management—two major themes of this report.

In its simplest sense, *sustainable* use of a resource means that the resource can be used indefinitely. But even a depleted resource can be used indefinitely at an undesirably low level, and perhaps with undesirable consequences. Therefore, by *sustainable fishing*, the committee means fishing activities that do not cause or lead to undesirable changes in biological and economic productivity, biological diversity, or ecosystem structure and functioning from one human generation to the next. Fishing is sustainable when it can be conducted over the long term at an acceptable level of biological and economic productivity[3] without leading to ecological changes that foreclose options for future generations. The desired levels of biological and economic productivity are in part societal decisions, but it is clear that both could be greater than they are today. In many cases, of course, sustainable fishing implies a need to rebuild populations of exploited species and to promote recovery of ecosystems from effects of overexploitation. *Ecosystem-based management* is an approach that takes major ecosystem components and services—both structural and functional—into account in managing fisheries. It values habitat, embraces a multispecies perspective, and is committed to understanding ecosystem processes. Its goal is to achieve sustainability by appropriate fishery management.

Humans are components of the ecosystems they inhabit and use. Their actions on land and in the oceans measurably affect ecosystems, and changes in

---

[3]Economic productivity means the generation of net economic benefits or profits.

ecosystems affect humans. Thus, sustainability of fisheries at an acceptable level of productivity and of the ecosystems they depend on requires a much broader understanding of appropriate and effective management than has been encompassed by traditional, single-species fishery management.

THE STATUS OF MARINE FISHERIES

Marine-capture fisheries include commercial, recreational, subsistence, and various small-scale fisheries, with total landings dominated by the commercial sector. In addition to recent reported annual landings of about 84 million t of marine animals (including fish, molluscs, crustaceans, and some other species), marine plants (seaweeds) also are used for food, as well as some marine mammals and turtles. Fishing as a source of food and revenue in less-industrialized countries, traditionally important, has become even more important recently and accounted for 65 percent of the world's catch in 1993.

In addition to fish and invertebrates that were caught and landed, approximately 27 million t of nontarget animals (bycatch) were discarded each year in the early 1990s (discards were probably less in the late 1990s). Furthermore, fishing causes mortality that is never observed because of illegal fishing, animals that die after escaping from fishing gear, or animals that are killed by abandoned fishing gear. Thus, the biomass of fish and invertebrates killed by ocean fishing (not including aquaculture) probably exceeds 110 million t per year.

Various estimates have been made of the total productivity of ocean ecosystems and the maximum long-term potential catch of marine animals. Many of the latter estimates are near 100 million t per year, suggesting that the current annual landings of 84 million t plus unreported mortality are near the maximum sustainable. However, considering species interactions, variations in the ability of individual species to withstand fishing mortality, global overfishing, and ecosystem degradation, it is possible that, under present management and fishery practices, the current catch cannot be exceeded or perhaps even continued on a sustainable basis. Considering individual stocks, about 30 percent globally are overfished,[4] depleted, or recovering, and 44 percent are being fished at or near the maximum long-term potential catch rate.

In the United States, commercial marine fishery landings in 1996 were 4.5 million t, worth $3.5 billion (exvessel value, the value of first sales from a vessel). The total economic contribution of recreational and commercial fishing were each approximately $20 billion per year. However, approximately 33 percent of stocks that commercial and recreational fishers depend on were over-

---

[4]By *overfishing* the committee means fishing at an intensity great enough to reduce fish populations below the size at which they could provide the maximum long-term potential (sustainable) yield (see Chapter 2), or at an intensity great enough to prevent their recovery to that size. As described in this report, it follows that overfishing is a function of population size.

fished or depleted in 1994, while 49 percent were fished at or near the level where they could yield the maximum long-term potential catch. In 1994, only about 2 percent by weight of total marine landings were from recreational fishing, but for several species recreational landings exceeded commercial landings.

FISHING AND MARINE ECOSYSTEMS

Fishing and ecosystems interact, and both are affected by environmental changes and other human activities. Fishing obviously has direct effects on fished stocks. It can alter abundance, age and size structure, sex ratio, genetic structure of fished populations, and species composition of marine communities. Many important commercial species are at high trophic levels (they eat other fishes), and their removal can have especially large effects on ecosystems, perhaps out of proportion to their abundance or biomass. Fishing can also affect habitats, most notably by destroying and disturbing bottom topography and the associated benthic communities. Large-scale mariculture activities (farming of fish, shrimp, and other marine organisms)—especially if they are poorly managed—also can affect marine ecosystems through damage to coastal wetlands and nearshore ecosystems associated with the construction of shore-based or nearshore facilities; through contamination of the water with food, antibiotics, and waste; and through the introduction of diseases and exotic genotypes.

Fishing has had significant effects on many marine ecosystems, including changes in productivity, biological diversity, and provision of ecosystem goods and services. For example, fishing has contributed to large changes in coral-reef ecosystems in the Caribbean, including the death of corals, and it has resulted in significant changes in community structure in the Bering, Barents, and Baltic seas, on Georges Bank, and elsewhere. In combination with environmental changes and other human activities that have led to the degradation of habitats, pollution, and the introduction of exotic species, fishing has had major effects in the Laurentian Great Lakes, San Francisco Bay, and Chesapeake Bay. It seems likely that, unless fishing and other activities are managed better, human effects on marine ecosystems will increase.

Long- and short-term environmental fluctuations have major effects on the abundances of marine organisms. Some well-known environmental fluctuations are those precipitated by El Niño events, which change the patterns of Pacific Ocean currents and affect global weather every few years. El Niños lead to the intrusion of warm water into high latitudes and major changes in the distribution and abundance of many species. Other environmental fluctuations affect marine areas at varying spatial scales and periods ranging from a few weeks to decades and perhaps centuries. Environmental changes can produce effects similar to those of fishing, and it is often difficult to distinguish them from the effects of fishing. Although they cannot be controlled directly, environmental fluctuations exert a fundamental influence on the behavior of marine ecosystems and must be taken into account by managers. To be sustainable, fishing and fishery management must be flexible and responsive to environmental changes as well as conservative of ecosystem components. Uncertainties about effects of environmental variability should not be used to excuse continued overfishing.

CONCLUSIONS AND RECOMMENDATIONS

Conclusions

Many populations and some species of marine organisms have been severely overfished. There are widespread problems of overcapacity: there is much more fishing power than needed to fish sustainably. Fishing affects other parts of the ecosystem in addition to the targeted species, and those effects are only now beginning to be understood and appreciated. Other human activities, such as coastal development, have adverse effects on marine ecosystems as well. The effects of these human activities, combined with ecosystem effects of fishing, may well be more serious in the long term than the direct effects of fishing on targeted species. Although societies have been concerned about the effects of fishing on particular populations and species for centuries, recent recognition of the ecosystem effects of fishing has resulted in part from research on ecosystem approaches and has led to calls for the adoption of ecosystem approaches to fishery management to achieve sustainability at a high level of productivity of fish and of ecosystem goods and services.

The committee concludes that a significant overall reduction in fishing mortality is the most comprehensive and immediate ecosystem-based approach to rebuilding and sustaining fisheries and marine ecosystems. The committee's specific recommendations, if implemented, would contribute to an overall reduction in fishing mortality in addition to providing other protective measures.

The committee recommends the adoption of an ecosystem-based approach for fishery management whose goal is to rebuild and sustain populations, species, biological communities, and marine ecosystems at high levels of productivity and biological diversity, so as not to jeopardize a wide range of goods and services from marine ecosystems, while providing food, revenue, and recreation for humans. An ecosystem-based approach that addresses overall fishing mortality will reinforce other approaches to substantially reduce overall fishing intensity. It will help produce the will to manage conservatively, which is required to rebuild depleted populations, reduce bycatch and discards, and reduce known and as-yet-unknown ecosystem effects. Although this approach will cause some economic and social pain at first, it need not result in reduced yields in the long term because rebuilding fish populations should offset a reduction in fishing intensity and increase the potential sustainable yields. Reducing fishing effort in the short term is necessary to achieve sustainable fishing. The options lie in deciding how and when to reduce effort so as to reduce economic and social disruption. The options, however, can be exercised only if decisions are made before the resources are depleted.

Adopting a successful ecosystem-based approach to managing fisheries is not easy, especially at a global or even continental scale. That is why the committee recommends incremental changes in various aspects of fishery management. The elements of this approach, many of which have been applied in single-species management, are outlined below. They include assignment of fishing rights or privileges to provide conservation incentives and reduce overcapacity, adoption of risk-averse precautionary approaches in the face of uncertainty, establishment of marine protected areas, and research.

When overfishing (including bycatch) has been effectively eliminated, other human activities will be the major threat to fisheries and marine ecosystems. Although those effects are not a major focus of this report, they cannot be totally separated from fishing, and mechanisms involving cross-sectoral institutional arrangements will be needed to protect fisheries and marine ecosystems.

Recommendations

The following are recommendations to achieve the broad goals and approach outlined above. Appropriate actions need careful consideration for each fishery and each ecosystem.

Conservative Single-Species Management

Managing single-species fisheries with an explicitly conservative, risk-averse approach should be a first step toward achieving sustainable marine fisheries. The precautionary approach should apply. A moderate level of exploitation might be a better goal for fisheries than full exploitation, because fishing at levels believed to provide the maximum long-term yield tends to lead to overexploitation. Many species are overfished and their productive potential is impaired, even without considering the ecosystem effects of fishing for them. Expanding fisheries to include previously unfished or lightly fished species, such as deep-sea species, is unlikely to lead to large, sustainable increases in marine capture fisheries. Therefore, the committee recommends that management agencies adopt regulations and policies that strongly favor conservative and precautionary management and that penalize overfishing, as called for in the Magnuson-Stevens Fishery Conservation and Management Act of 1976 and the 1996 amendments to that act, often referred to as the Sustainable Fisheries Act of 1996.

As described in Chapter 5, the committee's recommendation for more conservative and precautionary management requires that the concept of maximum sustainable yield be interpreted in a broader ecosystem context to take account of species interactions, environmental changes, an array of ecosystem goods and services, and scientific uncertainty. This step, although important, will not by itself sustain marine fisheries and ecosystems at high levels of productivity.

Incorporating Ecosystem Considerations Into Management

Fishery management should take account of known and probable goods and services of marine ecosystems that are potentially jeopardized by fishing. The aim is to sustain the capacity of ecosystems to produce goods and services at local to global scales and to provide equitable consideration of the rights and needs of all beneficiaries and users of ecosystem goods and services.

Dealing with Uncertainty

Fisheries are managed in an arena of uncertainty that includes an incomplete understanding of and ability to predict fish population dynamics, interactions among species, effects of environmental factors on fish populations, and effects of human actions. Therefore, successful fishery management must incorporate and deal with uncertainties and errors. The committee recommends the adoption of a precautionary approach in cases of uncertainty. Management should be risk-averse. Although research and better information can reduce uncertainty to a degree, they can never eliminate it.

Many of the problems that fishery managers face are issues concerning long-term versus short-term goals and benefits. Uncertainty in stock assessments and in future allocations of those stocks has led to an emphasis on short-term benefits at the expense of long-term solutions. Uncertainties over shares when allocations allow open competition can compel individuals to adopt a short-term horizon for decisions related to fishing effort and investment. Management incentives and institutional structures must counteract these responses to uncertainty that jeopardize sustainability. This is especially true when stock assessments are uncertain, which makes it harder for managers to hold the line on conservation.

Reducing Excess Fishing Capacity and Assignment of Rights

Excess fishing capacity (fishing capacity is the ability to catch fish or fishing power) and overcapitalization (capitalization, related to capacity, is the amount of capital invested in fishing vessels and gear) reduce the economic efficiency of fisheries and usually are associated with overfishing. Substantial global reductions in fishing capacity are of the highest priority to help to reduce overfishing and to deal with uncertainty and unexpected events in fisheries. Overcapacity is difficult to manage directly, and usually evolves in management regimes that encourage unrestricted competition for limited fishery resources. Consequently, managers and policy makers should focus on developing or encouraging socio-economic and other management incentives that discourage overcapacity and that reward conservative and efficient use of marine resources and their ecosystems.

At the core of today's overcapacity problem is the lack of, or ineffective, definition and assignment of rights in most fisheries. In addition, subsidies that circumvent market forces have contributed significantly to the overcapacity problem in many fisheries. Therefore, the committee recommends for many fisheries a management approach that includes the development and use of methods of allocation of exclusive shares of the fish resource or privileges and responsibilities (as opposed to open competition) and the elimination of subsidies that encourage overcapacity. A flexible and adaptive approach is essential, and careful attention must be given to equity issues associated with such approaches. The committee recommends experimental approaches to community-based fishery management, including the development of virtual communities. This would include research into the establishment of management groups in which participation is based on shared interests in a fishery and its associated ecosystem, with diminished emphasis on where participants live or their direct financial interests.

Marine Protected Areas

Where they have been used, marine protected areas—where fishing is prohibited—have often been effective in protecting and rebuilding ecosystems and populations of many (but not all) marine species. They often also lead to increases in the numbers of fish and other species in nearby waters. Importantly, they can provide a buffer against uncertainty, including management errors. Permanent marine protected areas should be established in appropriate locations adjacent to all the U.S. coasts. It will be important to include highly productive areas—that is, areas in which fishing is good or once was—if this management approach is to produce the greatest benefits.

Protected areas will make the most effective contribution to the management of species and ecosystems when they are integrated into management plans that cover the full life cycles and geographic ranges of the species involved. Smaller, fixed protected areas will be most effective for species with life stages that are spent in close association with fixed topography, such as reefs, banks, or canyons. For other species, the degree of effectiveness of protected areas will be related to the importance of fixed topography in various stages of their lives. Wholly or largely pelagic species move according to ocean currents or other factors that are not necessarily related to fixed topographic structures and are thus likely to benefit less from small protected areas.

The design and implementation of marine protected areas should involve fishers to ensure that they believe the resulting systems will protect their long-term interests and to improve operational integrity. Because attempts to develop marine protected areas in the United States have been strongly opposed by some fishers, the broad involvement of users is a key strategy. Current theory and experience make clear that marine protected areas must be established over a significant portion of the fishing grounds to have significant benefits. Recent calls for protecting 20 percent of potential fishing areas provide a worthwhile reference point for future consideration, and emphasize the importance of greatly expanding the areas currently protected.

Marine protected areas are not alternatives to other techniques of fishery management and to the other recommendations in this report. They should be considered as only one of a suite of important ecosystem approaches to achieve sustainable fisheries and protect marine ecosystems. For marine protected areas to be most successful as fishery-management tools, their intended purposes must be clearly defined.

Bycatch and Discards

Bycatch and discards add to fishing mortality and should be considered as part of fishing activities rather than only as side effects. Estimates of bycatch should be incorporated into fishery-management plans and should be taken into account in setting fishing quotas and in understanding and managing fishing to protect ecosystems and nonfished ecosystem components. Reducing fishing intensity on target species can reduce bycatch, often with no long-term reduction in sustainable yield. In some cases, technological developments and careful selection of fishing gear (e.g., bycatch-reduction devices) can be effective in reducing bycatch, and those options should be encouraged, developed, and required where appropriate. More information is needed on discards and on bycatch and their fate (i.e., whether bycatch is retained or discarded and whether discards survive or die).

Institutions

Effective fishery management requires structures that incorporate diverse views without being compromised by endless negotiations or conflicts of interest. The committee recommends developing institutional structures that promote

- effective and equitable reduction of excess capacity,
- sustainable catches of targeted species,
- expansion of the focus of fishery management to include all sources of environmental degradation that affect fisheries,
- consideration of the effects of fishing on ecosystems,
- development and implementation of effective monitoring and enforcement, and
- the collection and exchange of vital data.

To achieve these goals, the spatial and temporal scales at which the institutional structures operate should better match those of important processes that affect fisheries. Participation in management should be extended to all parties with significant interests in marine ecosystems that contain exploited marine organisms. Institutions should allocate shares in or rights to fisheries, rather than allowing openly competitive allocations. The clear explication of management goals and objectives is a prerequisite to achieving effective and equitable management.

Information Needs

Better understanding is needed of the structure and functioning of marine ecosystems, including the role of habitat and the factors affecting stability and resilience. This includes attempting to understand mechanisms at lower levels of organization (i.e., populations and communities), long-term research and monitoring programs, development of models that incorporate unobserved fishing mortality and environmental variability (e.g., El Niño events) into fishery models, multispecies models, and trophic models. More research is also needed on the biological effects of fishing, such as the alteration of gene pools and population structures as a consequence of fishing. More research is needed on the conditions under which marine protected areas are most effective, and marine protected areas themselves should be used as research tools as well as for conservation.

More information is needed on the effects and effectiveness of various forms of rights-based management approaches and other management regimes, on the way people behave in response to different economic and social incentives, and on barriers to cooperation and sharing of information. The committee recommends research into the roles of communities in fisheries management, including the use of community-based quotas and other assignments of rights to communities, and explorations into the feasibility of granting management responsibilities to those engaged in a particular fishery, regardless of their geographical community ("virtual communities").

The need for more information should not be used as an excuse for inaction; that excuse has contributed significantly to current problems. Enough is known to begin taking action now.

1

Introduction

NATURE OF THE PROBLEM AND THE COMMITTEE'S APPROACH

Declining marine fishery catches—including fish, crustaceans, and molluscs—have been the subject of much recent attention in the media and the technical literature (e.g., Botsford et al. 1997, Merrett and Haedrich 1997). For example, Atlantic halibut (*Hippoglossus hippoglossus*), once common off New England, are now rare. Marbled rockcod (*Notothenia rossi*) have been seriously depleted in the Southern Ocean. Off New Zealand, orange roughy (*Hoplostethus atlanticus*) populations have been significantly reduced because until recently, they were fished faster than they could replace themselves. Declines in bluefin tuna (*Thunnus thynnus*) populations in the North Atlantic are the subject of controversy and concern. Some fisheries have been subject to severe curtailment and closure; in North America, most notably cod (*Gadus morhua*) off Newfoundland, groundfish off New England, and some salmon species in the Pacific Northwest. Atlantic salmon (*Salmo salar*) and American shad (*Alosa sapidissima*) have largely or completely disappeared from many rivers of the eastern United States. Walleye pollock (*Theragra chalcogramma*) have been effectively eliminated from an area of the Bering Sea that lies in international waters (the Donut Hole), and many coral-reef fish species in the Philippines have disappeared from commercial catches. International disputes over fishery resources sometimes have a militaristic character, as illustrated by the "cod war" between the United Kingdom and Iceland in the 1960s and the Canadian arrest of a Spanish trawler on the high seas in the 1990s.

Sometimes, populations not directly fished might be affected. For example, yellowfin tuna (*Thunnus albacares*) fisheries in the eastern tropical Pacific until recent decades were responsible for the killing of perhaps hundreds of thousands of dolphins each year. The overall yield of the world's marine capture fisheries has not grown much in recent years. Not all fisheries are in decline, however. Some fisheries have been conducted for many years without depleting the fished populations, and some previously depleted fisheries have recovered.

How serious is the problem of declining marine fishery catches? Are we seeing only local depletions, with no significant effects on global catches? Or are the localized declines symptomatic or omens of a larger problem? Are other components of marine ecosystems being seriously affected, perhaps even to the point of significantly affecting ecosystem structure and functioning? To what degree are observed changes caused by environmental fluctuations and to what degree are they caused by human activities, particularly fishing? If the problems are significant, as they appear to be or likely to become, can an ecosystem approach to fishery management help achieve sustainability of marine fisheries?

To address these and related questions, the Ocean Studies Board (OSB) of the National Research Council (NRC) established the Committee on Ecosystem Management for Sustainable Marine Fisheries. The committee was charged to "assess the current state of fisheries resources; the basis for success and failure in marine fisheries management (including the role of science); and the implications of fishery activities to ecosystem structure and function. Each activity [was to] be considered relative to sustaining populations of fish and other marine resources" (Statement of Task). The committee, composed of experts in ecology, fishery biology, fishery management, economics, and anthropology, included members from academe, government, the fishing industry, and a nongovernmental organization.

The committee met five times over the course of its study. Its second meeting included an international conference in Monterey, California, in February 1996. The committee based its deliberations in part on the papers presented at that meeting (see Appendix A), many of which were published in a special issue of *Ecological Applications* (Vol. 8 Supplement, 1998). In addition, the committee reviewed a great deal of published literature, and the members contributed their own expertise and experience.

CONTEXT

Fishery management and this report need to be seen in their broader contexts. The world's human population continues to grow, and thus the demand for food—including seafood—continues to grow as well. Marine capture fisheries yielded a total of 84 million metric tons in 1995, by far the largest contributor to the 14 kg of fish available as food per person in 1995 (FAO 1997a).

Fisheries are a large international business. In 1996, first-sale revenues from fishery products (including aquaculture and both marine and freshwater production) were worth about $U.S. 100 billion (FAO 1996a); in 1992 they provided approximately 19 percent of the total human consumption of animal protein

(FAO 1992). Fisheries provided direct and indirect employment for about 200 million people (Garcia and Newton 1997). In addition, recreational fisheries account for large direct and indirect expenditures, especially in North America, Sweden, Australasia, and elsewhere.

The industrialization occurring in many parts of the world increases the need for foreign currency, and one way to get that currency is to sell seafood. Net fisheries exports in developing countries were worth $U.S. 16 billion in 1994 (FAO 1997a), more than the exports of coffee, bananas, rice, and many other commodities (FAO 1997b). As various fished populations decline, prices can rise. All these factors can increase fishing pressure. Recreational fishing also accounts for large landings, especially for a few game species in North America. In addition, the growing human population, increasingly industrialized, affects the terrestrial and marine environments in many ways, some of which might increase the fisheries catches, but many do not. As described in Chapter 3, destruction of spawning and nursery habitats, disruption of food webs, nutrient and other chemical pollution, and sedimentation can all adversely affect fisheries. Thus, the possibility of increasing global marine fishery catches by increasing fishing effort seems increasingly remote. Indeed, it appears likely that we will need to *reduce* effort to sustain the current catch rate.

Society's way of looking at marine environments also has changed. International agreements reached over the past two decades increasingly recognize the importance of marine ecosystems, the need to sustain them, and the vital links between terrestrial and marine systems. People are increasingly aware of the effects of fishing on other ecosystem components such as dolphins, turtles, birds, many invertebrates, and others. The value of the goods and services provided by ecosystems is increasingly being recognized. For example, Costanza et al. (1997) estimated the world's ecosystem services at $U.S. 16 trillion to 54 trillion per year, with more than half that value derived from marine ecosystems.

Marine ecosystem services operate over a wide range of spatial and temporal scales. They range from climate regulation, operating at the global scale, to more local services such as the provision of habitat for nursery or spawning grounds or the protection of shorelines from battering by waves. Kelp forests, mangroves, coastal wetlands, and coral reefs provide habitats and protect shorelines from erosion. Estuaries and mangroves trap sediment, thus protecting downstream ecosystems such as coral reefs. Microbes in sediments can detoxify many pollutants, and others are sequestered by the sediments. In addition to food, goods include chemicals like algin and carrageenan from seaweeds. Other services, such as biogeochemical cycling in the oceans, are obviously important, although we know little of their details. Ecosystem values include many nonmarket ones, such as opportunities for recreation and aesthetic enjoyment, as well as the ecosystem services described above (Daily 1997). Some even accrue from merely knowing that the ecosystems and their components exist, even though most people will never see them. There is increasing recognition that sustaining fishery yields will require sustaining the ecosystems that produce the fish. Moreover, there are compelling reasons beyond fishing to sustain marine ecosystems. Thus, this report is about sustaining ecosystems rather than sustaining only fishery catches.

SUSTAINABILITY AND ECOSYSTEM-BASED MANAGEMENT

In establishing the Committee on Ecosystem Management for Sustainable Marine Fisheries, the NRC and OSB recognized that the challenge of achieving sustainable fisheries is greater than the challenge of achieving a sustainable catch of each important commercial species over the next decade or two—a daunting task in itself. Because there is evidence (reviewed in Chapter 3) that the ecosystems in which exploited species live can be affected by fishing and that changes in those ecosystems can affect the exploited species, it was apparent to the committee that achieving sustainable fisheries would require a broader approach to fishery management than has been common: an ecosystem-based approach. Addressing this question, however, requires some agreement on the meaning of two terms, *sustainability* and *ecosystem-based management*.

Sustainability

Sustainability is an important idea, although it is hard to define precisely. The central idea is that a resource is used in such a way that it is not depleted or permanently damaged. In other words, use of the resource can be continued indefinitely. Defining the limits of "depletion," "permanent damage," "indefinitely," and related terms is difficult (Norgaard 1994), in part because the physical and biological components of the world keep changing. Nonetheless, the committee agrees that the concept provides a useful goal. To implement the goal, it is critical to understand the distinction between maintaining a particular catch rate over a short period as opposed to maintaining the continued productivity of the ecosystem, which is required to produce the species of interest (and others). Therefore, for the purposes of this report, the committee defines *sustainable fisheries* to be fishing activities that do not cause or lead to undesirable changes in biological and economic productivity, biological diversity, or ecosystem structure and functioning from one human generation to the next (see Lubchenco et al. 1991); sustainable fishing does not lead to ecological changes that foreclose options for future generations.

Sustainable fishing can take place at different levels of productivity and abundance of many species: in many places, fish populations have been over-fished with little change in abundance for long periods. In addition, there are many cases in which the biological and economic productivity of fish populations, ecosystems, and fisheries would be enhanced if the fish populations were allowed to rebuild; this would represent a change from one year to the next. In those cases, the goal the committee is referring to includes rebuilding of those populations and at least some recovery of the ecosystems. The committee recognizes that environmental fluctuations and minor or short-term changes caused by fishing will continue to occur. But the goal implied by the definition above seems clear, as has a failure to achieve it in many cases thus far.

Ecosystem-Based Management

Although *ecosystem management* has many definitions, *sustainability* is a central part of most of them (Christensen et al. 1996, NRC 1996a). We cannot really manage ecosystems per se; instead, it is human activities that are managed. The committee therefore uses the term *ecosystem-based management.*

To the degree that it is successful, management that focuses on ecosystem structure and functioning will also improve the sustainability of fisheries. Similarly, making fishing sustainable will help to sustain marine ecosystems, albeit altered from their pre-exploitation states. But current scientific knowledge does not permit us to manage large marine ecosystems comprehensively and reliably. Therefore, by *ecosystem-based management* for fisheries, this committee means an approach that seriously takes all major ecosystem components and services— both structural and functional—into account in managing fisheries and one that is committed to understanding larger ecosystem processes for the goal of achieving sustainability in fishery management.

Concepts of ecosystem management and sustainability are not new, although their explicit incorporation into many management goals is fairly recent. For example, Kurien (1998) described traditional Asian coastal proverbs that used to guide traditional fishing activities. Two of them are closely related to the ideas of sustainability and ecosystem management: *The wealth of the sea belongs to the dead, the living, and those yet to be born* and *Where there is water there is fish; if we take care of the water, the fish will take care of us.*

Humans as Ecosystem Components

That humans are components of the ecosystems they inhabit and use seems obvious, but it is often overlooked. Too often, managers divide the world into "the ecosystem" and "the users of the ecosystem." Such a division is artificial and can lead to the absurd conclusion that the best way to achieve sustainability of an ecosystem is to keep people out of it. Humans are integral parts of the ecosystems they inhabit and use, and their actions on land and in the oceans affect the ecosystems, just as changes in those ecosystems affect humans. Sustainability applies to them as well as to other ecosystem components.

EVOLUTION OF VIEWS OF FISHERY MANAGEMENT

Fishing is an old activity, and concern about its effects also is old. As early as the fourteenth century, people in England were worried about fishing with a *wondyrchaum* (a fine-meshed trawl), which they feared was killing enormous numbers of small fish (March 1970). By 1716, minimum mesh sizes and minimum size limits for various fish species were in effect. Many forms of fishery management have evolved based on traditional knowledge gained by fishing peoples. They included cultural practices, community agreements, and government controls. But in general, and despite early regulations, management of marine fisheries was minimal before the middle of the twentieth century, with a few notable exceptions such as the International Pacific Halibut Commission (Bell 1978).

Although there has long been concern over the effects of fishing, that concern—like fishing itself—was largely confined to coastal waters until recently. The famous scientist Thomas Huxley expressed his opinion in 1883 that probably "all the great sea-fisheries are inexhaustible" (Smith 1994), although some great sea-fisheries were already depleted by then. The Atlantic halibut fishery had collapsed by the early 1880s (Goode and Collins 1887) and has not yet recovered. And the U.S. Fish Commission had been established a dozen years earlier (1871) to find the causes of declining New England fisheries.

By 1919, W.F. Thompson (1919) recognized that sea fisheries were not inexhaustible, but he identified a difficulty that remains critical to this day: "Proof that seeks to modify the ways of commerce or of sport must be overwhelming." He developed research to concentrate on what "is necessary to the perpetuation and prosperity of the fishery." To protect the Pacific halibut (*Hippoglossus stenolepis*) from the fate of the Atlantic halibut, the International (United States and Canada) Pacific Halibut Commission was established in 1924 with "a competent man as director of investigations": W.F. Thompson (Babcock et al. 1928). While and before Thompson was investigating the North Pacific halibut fishery, Heincke was developing catch-curve analyses as a basis for proposed minimum size limits for North Sea plaice (*Pleuronectes platessa*) and demonstrating racial differences in North Sea herring (Smith 1994). The International Council for the Exploration of the Seas was formed in 1902, motivated by the desire to understand and predict fluctuations in fish stocks (Smith 1994).

North Sea fish increased in size and numbers during both World Wars I and II, because naval action forced a reduction in fishing. That phenomenon led to wider recognition that fishing did affect fish stocks and that the effects were at least partly reversible. Those observations led Graham (1935, 1943) to explain how increases in fishing power allowed catches to be maintained or increased even when fish stocks declined, leading to a waste of time and money by comparison with fishing at a maximum sustained yield.[1] Gordon (1954) provided the economic theory to support Graham's observations.

In the 1950s and 1960s, scientific methods of stock assessment and estimation of fishery yields were developed based on growth rates and age composition of the catch (e.g., Ricker 1954, 1958; Beverton and Holt 1957, Murphy 1965, Gulland 1965). At the same time, fishing vessels based many hundreds or thousands of miles from fishing grounds ("distant-water fleets") changed the face of fishery management, leading to exclusive economic zones, international treaties, international fishery-management bodies, and continued international disputes about fishing. One of those international bodies, the U.S-Canadian International Commission for Northwest Atlantic Fisheries, implemented an early attempt at incorporating ecosystem management into fisheries. When it set catch limits for a particular species, it took into account the expected bycatch of that species in fisheries for other species. It also concluded that the total sustainable catch rate for all species was less than the sum of sustainable catch rates for the individual species.

As described in Chapter 4, many factors have reduced the effectiveness of management and prevented the adoption of scientific advice. In a few cases the advice itself was not correct. Progress in developing better scientific perspectives for management and more equitable and effective ways of implementing management goals continues, but it has been slow. In some cases the obstacles seem to be overwhelming. Significant societal, commercial, and governmental economic incentives and pressures often lead to unsustainable fishing practices. It is difficult, for example, for a poor, hungry family to stop or even reduce fishing if its basic food needs depend on fishing and it has no resources with which to supplement reduced food from fishing. In addition, most such families live in countries whose governments have limited resources to apply to the problem. In wealthier societies, economic pressures can be as great as the basic need for food. But it is clear that the long-term costs of overexploitation of fishery resources are even greater than the short-term costs of reducing catches. Options for achieving sustainable fishing are the topic of this report.

Terminology

As views of fishery management evolve, the importance of terms becomes apparent. For example, it is common to refer to fishing as "harvesting the resource." But the term *harvest* usually includes the idea that an investment has been made in a crop, which is not true of most marine-capture fisheries. Another

---

[1]*Maximum sustainable yield* (MSY) is the largest average catch that can be captured from a stock under existing environmental conditions (NRC 1998a). *Maximum economic yield* (MEY) is the level of catch that provides the maximum net economic benefits or profits to society (Clark 1990, NMFS 1996b). MEY is usually less than MSY. Economic factors that influence fishing are discussed in Chapter 4.

term often used is *underutilized*, which implies that greater exploitation of a particular fish stock is desirable. This term probably reflects one purpose of many fishery-management institutions—that is to promote fisheries. In this report those terms and some others are avoided; in general, the committee has sought to use the terms *catch* or *landings* instead of *harvest* and *lightly fished* or *subject to low fishing mortality* rather than *underutilized*. The term *stock* might be linked by analogy to stock of a commodity. It also is difficult to define biologically, and so the committee has used the term *population* where possible. It is extremely difficult to find terms that do not express any societal values. Thoughtful evaluation of the values implicit in many terms and replacement of some of them could be an important part of achieving sustainable fishery management.

REPORT ORGANIZATION

Chapter 2 sets the stage by reviewing information on the state of marine fisheries. The committee attempted to be comprehensive although not exhaustive. Thus, enough information is presented for a general overview, but not all relevant examples are discussed. In Chapter 3 the effects on ecosystems of fishing and of environmental changes are discussed. Chapter 4 describes what is known about the factors that contribute to the conditions described in chapters 2 and 3. These include incomplete scientific information, scientific errors, failure to heed scientific advice, conflicting and unresolved goals and values, failure to consider all ecosystem components, institutional failures, social and economic incentives that do not favor sustainable resource use, changes in perspectives, new information, and perhaps a failure to recognize the limits of science.

Chapter 5 applies the lessons learned in Chapter 4. The discussion covers the potential usefulness and practical application of an ecosystem-based approach, alternative institutional structures, social and economic arrangements that hold promise for improving sustainable fishing, marine protected areas, scientific questions, and research needs. The emerging recognition that marine ecosystems have values in addition to their production of food and the importance of that recognition in developing ecosystem-based approaches to sustainable fishery management are also discussed. Finally, Chapter 6 contains the committee's conclusions and recommendations.

2

Current Status of Marine Fisheries

In general, considered on a single-species basis, many marine fisheries are fully exploited or overexploited, while relatively few seem to have the potential for increased exploitation.[1] In general, this is true both for the United States and globally, especially in estuarine, nearshore, and continental-shelf fisheries, which produce approximately 75 percent of the world's fish catches (Pauly and Christensen 1995).

The primary source of global information about the condition of fisheries is the Food and Agricultural Organization (FAO) of the United Nations. In the United States the task of carrying out assessments has been primarily the responsibility of the National Marine Fisheries Service (NMFS) of the National Oceanic and Atmospheric Administration, U.S. Department of Commerce. Regional fishery management councils (established under the authority of the Magnuson-Stevens Fishery Conservation and Management Act of 1976) also are involved in the assessment of fish stocks in federal waters. Interstate fishery commissions in the Atlantic, Gulf, and Pacific regions work with states and the NMFS to conduct assessments of migratory fish stocks in state waters. Although assessments and statistics from FAO and NMFS provide only an imperfect characterization of the status of global and U.S. fisheries, the assessments—corroborated by many kinds

---

[1]Estimates of utilization, according to the National Marine Fisheries Service's terminology (1993, 1996b), are based on the concept of *long-term potential yield* (LTPY), the maximum long-term (or sustainable) average that can be maintained with conscientious stewardship through regulating total catch. A fishery resource is *fully utilized* when the current fishing effort is about equal to the amount needed to achieve LTPY. If the effort is greater than that, the resource is considered to be *overutilized*; if the fishing effort is less, the stock is considered to be *underutilized*. As noted in Chapter 1, this terminology should be reevaluated.

of evidence—appear to provide a reasonably accurate description of the overall picture.

GLOBAL OVERVIEW

Fishing is an important source of food, recreation, community development, wealth, and cultural values in many countries. Although thousands of freshwater and marine fish and shellfish species are used globally, a relatively small number of these species provide the major fraction of the global marine catch. The 10 marine species that provided the greatest catch in 1993 accounted for 35 percent of the commercial marine catch (Figure 2-1, FAO 1996b) and the top 20 species accounted for 46 percent of the global marine catch. Marine fish production is shown in Figure 2-2 and total fish production in Figure 2-3. In addition to the animals mentioned, marine algae (seaweeds) are extensively harvested in many parts of the globe (Abbott and Norris 1985, Akatsuka 1990, Akatsuka 1994, Santelices 1989, Tseng 1984).

Recent estimates indicate that the global first-sale revenues from fishery products are approximately $U.S. 95 billion annually and that fishery products account for about 20 percent of the animal protein consumed by humans (FAO 1995b). Fisheries provide direct and indirect employment to about 200 million people worldwide (Garcia and Newton 1997). Fisheries are especially important in developing countries, which increased their proportion of global catch from about 40 to 65 percent from 1973 to 1993. The net value of fishery products exported from developing countries totaled $16 billion in 1994 (FAO 1997a), greater than the exports of coffee, bananas, rubber, tea, rice, and many other commodities that developing countries have traditionally relied on for foreign exchange (FAO 1997b).

Global marine fish production increased at an average rate of about 3.6 percent per year from 1950 to 1995, from about 18 million to about 91 million metric tons,[2] including mariculture production (Figure 2-2, FAO 1997a). In the same period, the world's population increased from 2.5 billion to 5.7 billion people (U.S. Census Bureau 1998), an average annual increase of 1.8 percent. In 1995 total fish production (both freshwater and marine, both through culture and through fishing) was approximately 112 million t (Figure 2-3), of which marine landings accounted for approximately 84 million t. In 1996 the total production reached approximately 116 million t; the increase was due mainly to an increase in freshwater aquaculture production, mainly in China (FAO 1997c). The supply of fish and fish products for human consumption (including freshwater fish and aquaculture products) reached roughly 14 kg per person annually in 1995 (FAO 1997a). By 1995, mariculture accounted for 6.7 million t, 7.4 percent of the total global marine fish yield; freshwater aquaculture provided 14.6 million t (FAO

---

[2]One metric ton, or tonne (t), is 1,000 kg and equals approximately 2,205 pounds.

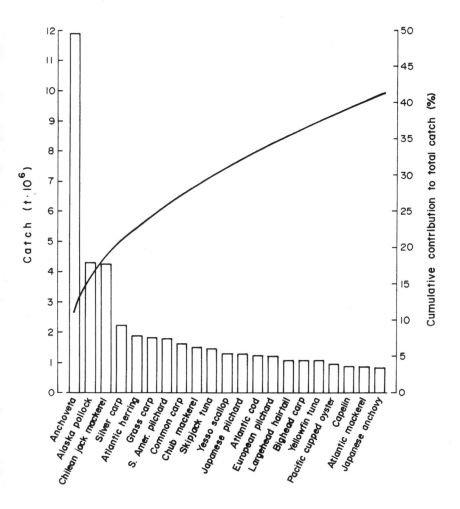

FIGURE 2-1   Total production by principal (mainly marine) species in 1994. Production includes mariculture and aquaculture, but represents catch for most species. Source: Redrawn from FAO (1996b).

1997a). Approximately 31.5 million t (28 percent) of world fish production was used for animal feed—including feed for mariculture—and other products that do not contribute directly to the human food supply in 1995.

In addition to fish that are caught and processed, a substantial number of fish and other organisms are caught and discarded—usually dead—at sea. Discards are a result of bycatch, which results because fishing gear and methods are not selective enough to catch only the target species, and of highgrading, the discarding of smaller or less desirable fish in favor of larger or more desirable fish that are caught later. Some bycatch is retained, but the remainder is discarded when

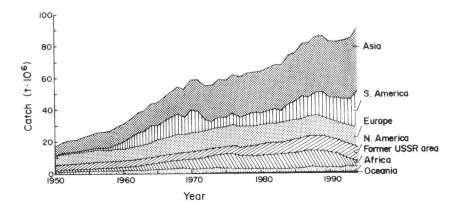

FIGURE 2-2   Total marine fishery production in 1994. Production includes mariculture but represents catch for most species. Source: Redrawn from FAO (1996b).

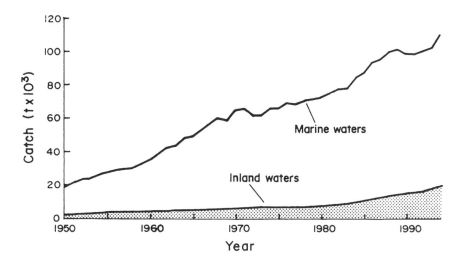

FIGURE 2-3   Total world fishery production, including freshwater and marine, for 1994. Production includes aquaculture but represents catch for most marine species. Source: Redrawn from FAO (1996b).

the species, size, quality, or condition of the fish reduce their value, or when fishery management regulations prohibit their retention. Alverson et al. (1994) estimated that commercial marine fisheries around the world discarded an estimated 27 million t of nontarget animals in the early 1990s, an additional biomass about one-third as large as total landings.

In addition to fish mortality caused by landings and discards, fishing can cause additional *unaccounted* mortality. Potential causes of unaccounted mortality include illegal or misreported landings; escapement or avoidance mortality that occurs when fish are injured by fishing gear but are not captured; and ghost fishing mortality, caused by lost gear (e.g., traps and gillnets) that continues to catch fish. The magnitude of unaccounted mortality is unknown but may be high for some fisheries. For example, the Scottish Fishermen's Federation estimated illegal or misreported landings, probably of groundfish and crustaceans, to be 100 to 200 percent of the reported catch (ICES 1995). Myers et al. (1997) concluded that discards of young undersized fish were an important reason that the fishing mortality of northern cod off Canada's maritime provinces was consistently underestimated in the 1980s, leading to the overfishing that caused the collapse of the fishery (see discussion of this case at the end of this chapter). Although these examples are not necessarily representative of all fisheries, they show that the total mortality resulting from fishing can easily be underestimated.

The FAO (1994a) and Garcia and Newton (1997) have concluded that the relatively stable catches of the early 1990s indicated that capture fisheries are near, or have reached, their sustainable limit based on existing fishing techniques and market systems. The increase in catch between 1950 and 1994 shown in Figure 2-2 occurred because of steadily increasing demand for fishery products, resulting in increased fishing capacity and effort. Fisheries were developed or expanded on formerly less-exploited or unexploited species and populations. While global assessments indicate that there are still opportunities to expand some fisheries, most are fully exploited or beyond, based on single-species considerations (Garcia and Newton 1997). Some have been so depleted that they are producing much less than their long-term potentials. Global fishing capacity is much greater than needed for sustainable marine fisheries, again based on single-species considerations.

More than 25 years ago, Gulland (1972) estimated that the potential sustainable yield from traditional fishery resource species (excluding Antarctic krill and oceanic mesopelagic fishes) was about 100 million t, with a practical limit (due to imperfect management and multispecies interactions) of about 80 million t, similar to catches that have been achieved in recent years and to recent FAO assessments.

Maximum potential global marine fisheries yield has also been estimated by Schaefer (200 million t, 1965), Ryther (100 million t, 1969), Idyll (400-700 million t, 1978), Houde and Rutherford (more than 300 million t from all marine ecosystems, 1993), and others. Based on those estimates, one might conclude that we have not yet reached maximum global fisheries yield. However, most of the highest estimates—admittedly upper limits in some cases—were made using unrealistic assumptions about food-web structure (Pauly 1996), effects of bycatch, feedback effects of fishing on other fish populations and marine ecosystems, and the technical and economic feasibility of new fisheries.

Limits to Global Production

Three kinds of information suggest that marine fish catch is near, at, or above its maximum sustainable level: estimates of theoretical limits imposed by available primary production, information on the degree of utilization of fish populations, and information on the catch per ton of fishing vessel.

Food-Web Limitations

The capacity of the ocean to produce fish is limited in part by the amount of marine phytoplankton produced annually. Fishery landings tend to be higher from ecosystems with higher levels of primary production, especially marine areas characterized by fronts, convergence, and upwelling areas. Satellite and in situ measurements of phytoplankton concentrations and in situ measurements of nutrients, water temperature, irradiance, and primary production allow estimates of the primary production of the global ocean, as well as regional estimates. An upper limit to the ocean's potential fisheries yield has been estimated many times by applying knowledge of the amount and location of global primary production, trophic level of the catch, and the transfer efficiency of biomass among trophic levels.

Pauly and Christensen (1995) used global catch data—which they divided according to trophic level—to estimate the flow of carbon up through the trophic levels of global marine ecosystems. They estimated that the transfer efficiency between trophic levels was about 10 percent, and concluded that about one-quarter to one-third of total primary production in coastal and continental shelf waters is needed to support recorded landings plus discards.

Houde and Rutherford (1993) used relationships between catches and primary production (Nixon 1988) and between fish production and primary production (Iverson 1990) to estimate a partitioned global fisheries production for estuaries, coastal zones, and upwelling areas. They estimated a total global fisheries production in those ecosystems of 543 million t, from which 111 million t might be removed as yield. (Note that Houde and Rutherford's estimate of the total potential yield of more than 300 million t was based on an estimated production of more than 1,300 million t in all marine ecosystems. They considered open-ocean production to be technologically difficult to use.) These estimates suggest that landings are near or beyond their sustainable limit, particularly if fish production lost as discards and unaccounted mortality are considered.

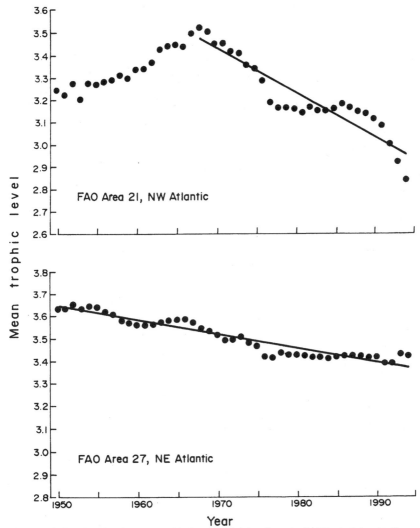

FIGURE 2-4   Trends of mean trophic level of fish landings in the North Atlantic. Source: Redrawn from Pauly et al. (1998).

Another line of evidence suggesting that global marine catch might not increase, even by fishing at progressively lower trophic levels, is provided by analyzing changes in the mean trophic level of marine fishery landings. Figure 2-4 shows the mean trophic level of global marine catches and of catches from FAO areas 21 (northwest Atlantic) and 27 (northeast Atlantic) based on species- or group-specific trophic levels taken from FAO fishery statistics and FishBase 97 (see www.fishbase.org), and described in detail by Pauly et al. (1998).

Figure 2-4 also shows significant declines in the average trophic level of fish catches from the 1950s for the northeast Atlantic and from the 1970s for the northwest Atlantic. This reflects a decrease in the proportion of long-lived carnivores in the catch relative to shorter-lived smaller pelagics and invertebrates. Fishing down the food web, while overfishing higher trophic forms, does not necessarily lead to increased total catches. As fishing takes animals lower in the food web, an increasing portion of the total catch may consist of animals for which there are no current markets or that are so diffuse that the cost of their capture does not warrant the expense (e.g., some large zooplankton species). In addition, the loss of predators (i.e., animals higher in the food web) can lead to an increase in competitors of the target species. The average trophic level of landed species can drop rapidly as catches of top predators or decline, as observed in most other FAO areas analyzed in this fashion (Pauly et al. 1998).

Degree of Fish-Stock Utilization

FAO periodically reports the degree of utilization of global fish stocks, classifying fisheries as underexploited,[3] moderately exploited, heavily to fully exploited, overexploited, depleted, and recovering. The largest number of fisheries (44 percent) are classified as heavily to fully exploited. Twenty-five percent of stocks have been fished beyond sustainable limits (overexploited, depleted, and recovering). For the United States during the period 1992-1994, the picture was similar despite slight differences in terminology: 12 percent of 275 stock groups were classified as underutilized, 34 percent as fully utilized, 23 percent as overutilized, and 31 percent were of unknown status (NMFS 1996a). Of the 191 stock groups whose status was known, 82 percent were fully utilized or overutilized. A U.S. example of a formerly overexploited and now recovered stock is striped bass (*Morone saxatilis*); Georges Bank haddock (*Melanogrammus aeglefinus*) represents a depleted stock.

Globally, some increase in exploitation might be possible for 32 percent of the landed species, but Garcia and Newton (1997) noted that, given past experience, heavily to fully exploited fisheries are likely candidates for future overfishing. This assertion is demonstrated by Alverson et al. (1994), who reported

---

[3]We have used FAO's terminology here, as we have used NMFS's similar terminology (*underutilized*) in quoting U.S. figures below. This does not constitute an endorsement of the terms by this committee (see Chapter 1). Clearly, the terms *underexploited* and *underutilized* imply a policy goal of full utilization, however that is defined. The terminology reflects a particular policy framework. One of the major arguments in this report is that aspects of the policy framework of our relationship to marine ecosystems need reexamination. The committee does not at present take a position on the desirability of the above terms but recommends that readers keep implied policy frameworks in mind.

(based on FAO data) that from 1980 to 1990 the number of overexploited fisheries increased by 250 percent, whereas the number of underexploited fisheries decreased by about 75 percent. Depleted species are being replaced in today's catches by species that were less heavily fished in the past.

For example, the Chilean Inca scad (*Trachurus murphyi*), Japanese pilchard (*Sardinops sagax melanosticus*), South American pilchard or Chilean sardine (*Sardinops sagax sagax*), and skipjack tuna (*Katsuwonus pelamis*) replaced chub mackerel (*Scomber japonicus*), Atlantic mackerel (*Scomber scombrus*), Atlantic cutlassfish (*Trichiurus lepturus*), and saithe (Atlantic pollock, *Pollachius virens*) in the top-10 species list between 1973 and 1993. In the United States, skates and dogfish have replaced more commercially valuable fish on Georges Bank. This process of depletion of one resource and replacement by another is limited by the number of potentially catchable and usable species; depleted species may not return to previous abundance levels. For example, Atlantic halibut (*Hipploglossus hippoglossus*) and spring-spawning Icelandic herring (*Clupea harengus*) have not recovered from overfishing, although they probably would if mortality were reduced (Myers et al. 1997). Many shark populations appear to be declining as well. For instance, Van der Elst (1979) described the impact of South African antishark nets on local populations of oceanic sharks, and the resultant and unanticipated decline in nearshore bony fishes. On a global scale, Manire and Gruber (1990) concluded that sharks were overfished by approximately 30 percent per year in U.S. waters; and cited domestic demand for shark meat; wasteful fisheries practices, especially discarded bycatch of sharks; irrational dread; and an increasing global demand for shark fins as major factors contributing to excess fishing mortality of sharks.

In addition, unexploited fish populations that are long lived and slow growing cannot support high exploitation rates, unlike populations of faster-growing, short-lived species. For example, the five species of the genus *Sebastes*, including the Pacific Ocean perch itself (*S. alutus*) and the northern (*S. polyspinus*), rougheye (*S. aleutianus*), sharpchin (*S. zacentrus*), and shortraker (*S. borealis*) rockfishes off the northwestern and Alaskan coasts of the United States and the coast of British Columbia, are all slow-growing and were severely overfished, although they have now largely recovered (NPFMC 1997, NMFS 1996b). The marbled rockcod (*Notothenia rossi*) in the Southern Ocean also has been severely overfished (Kock 1992).

Catch Per Ton of Fishing Vessel

Approximately 3.5 million vessels are engaged in fisheries worldwide; about two-thirds are small undecked vessels (FAO 1995b), but the total also includes about 24,000 high-seas fishing vessels of more than 500 gross tons (NMFS 1993). The gross tonnage of the world's fishing fleets (decked vessels only) increased by an average of 2.9 percent annually from 1970 to 1992. This rate of increase in gross tonnage exceeded the rate of increase in catch (1.8 percent annually) during the same period: the ratio of metric tons of fish caught per ton of fishing vessel decreased from 4.3 in 1970 to 3.0 in 1992. Assuming that additional fishing capacity has at least the same average fishing power per ton as the preexisting fleet—almost certainly true—and that the new vessels are used at least as much as the older ones—probably the case—the decreased ratio of catch to fishing tonnage provides further evidence that fish populations declined on average during this period. The discrepancy between fishing capacity and catch is even greater when one considers the increase in fishing power of vessels as a result of technological improvements. Based on an analysis conducted by Fitzpatrick (1995), the rate of increase in fishing power resulting from technological improvements has averaged 4.4 percent annually since 1965. Garcia and Newton (1997) fit a production model to global catch and gross tonnage data, adjusted for fishing power increases. Their analysis indicates that catch is higher than the maximum sustainable yield of world fisheries and that fishing capacity is too large to be economically efficient.

The decline in the per-ton catch rate of fishing vessels also indicates an economic problem, although lower catches have probably been partially offset by price increases. The economic problems also include the substantial debt service or depreciation of fishing vessels. Government subsidies have been used worldwide to increase employment and food supply. Subsidies have probably stimulated excess growth in the world's fishing fleet and must be a major factor in poor economic performance. They may amount to as much as $27 billion per year, although information about subsidies and how people and organizations react to them is not readily available, as discussed in Chapter 4. These problems, usually referred to as overcapitalization or excess fishing capacity, are discussed in more detail in chapters 4 and 5.

UNITED STATES OVERVIEW

Fishing Sectors

Marine fishing activities in the United States are divided among commercial, recreational,[4] subsistence,[5] and indigenous sectors. The balance of activity among these sectors depends on the areas and species fished and whether the comparison is made in terms of weight or number of fish landed or dollars injected into the U.S. economy. All sectors are subject to fishery management in the United States through the regional fishery management councils and, in some cases, through

---

[4]Larkin (1972) described recreational fishers as commercial fishers who are independently wealthy and subsidize their fishing from outside sources. He made the important point that there is considerable overlap between commercial and recreational fishing.

[5]Subsistence fisheries are most often carried out by indigenous peoples, but, especially in Alaska, other groups also conduct subsistence fisheries.

state and international agreements as well. Fisheries are important to the culture and social structure of their practitioners and can have a major economic impact, at least regionally.

Commercial Fisheries

The United States has the largest exclusive economic zone (EEZ) of any nation, covering about 11 million km^2. The United States was the fifth-largest fish producer in 1993, following China, Japan, Peru, and Chile (FAO 1995b). The first-sale value of U.S. commercial landings (4.47 million t[6]) in 1997 was estimated at $3.5 billion (NMFS 1998), with a direct contribution to the gross domestic product (GDP) of $20.2 billion. The United States is also one of the world's largest fish-trading nations, with a deficit of $4.6 billion in 1994 resulting from $12 billion in imports and $7.4 billion in exports (NMFS 1995a).

U.S. commercial landings were relatively stable at about 2 million t per year from 1935 until 1977, when the United States extended its jurisdiction over fisheries to 200 miles from the coast and increasingly excluded foreign vessels. At present, foreign fishing is not permitted in the U.S. EEZ, although in some cases—for example, menhaden (*Brevoortia tyrannus*) in the Gulf of Maine— foreign processor vessels receive catches from the U.S. EEZ. Since 1977, land- ings have more than doubled, to 4.47 million t in 1997 (NMFS 1998). The rapid rise in U.S. catch in the late 1980s was due primarily to the walleye pollock fishery that resulted from displacements of foreign vessels during the 1970s and into the 1980s (Figure 2-5). About half of the U.S. landings are from the fishing grounds off Alaska, primarily walleye pollock (*Theragra chalcogramma*), Pa- cific cod (*Gadus macrocephalus*), and various salmon (*Oncorhynchus*) species. As is true for most fishing nations, U.S. fishers are dependent on a small number of species, with almost 50 percent of the catch composed of walleye pollock from the Pacific Ocean and menhaden (*Brevoortia tyrannus* and *B. patronus*) from the Gulf of Mexico and Atlantic Ocean.

Recreational Fisheries

Recreational fishing also is important in the United States. Although the recreational catch is only about 2 percent as large as commercial landings for all species combined (90,000 t in 1994), there are more than 17 million marine recreational fishers, who in recent years made more than 66 million fishing trips per year, caught about 360 million fish, and spent $25.3 billion per year on

---

[6]This number includes the weight of the meat but not the shells of shellfish. FAO statistics usually include the weight of the shells also. When FAO reports landings for the United States (and other countries), it estimates shell weight and thus the weight is usually about 0.7 million t higher for U.S. landings than the weight given usually in U.S. publications (D. Sutherland, NMFS, personal commu- nication, 1998).

fishing-related activities (NMFS 1995a), comparable to the contribution to the GDP of commercial fisheries. For some fisheries in which both commercial and recreational fishers participate (e.g., summer flounder [*Paralichthys dentatus*] and bluefish [*Pomatomus saltatrix*]), the recreational catch is a significant portion or even a majority of the total (Table 2-1).

Recreational and commercial fishers often conflict over management goals and methods for various fisheries. In some cases, recreational fishers are effective at influencing policy, as for example recent restrictions they supported on the use of nets in coastal waters of various states (including a legislative ban on gillnets in Texas in 1988; California's Proposition 132, which banned net fishing starting in 1990; a Florida legislative ban on coastal nets that passed in 1993; and a Louisiana legislative restriction on nets passed in 1994). In other cases, they are not successful. The allocation of available marine fisheries resources between commercial and recreational sectors is a major issue for regional fishery management councils and in the political arena. Some of the disputes and the differences—and occasional agreements—between commercial and recreational fishers are described in almost every issue of *National Fisherman* and *Saltwater Sportsman*; for a discussion of net bans, for example, see the August 1996 issue of *National Fisherman*. The resolution of such disputes and allocation controversies is made more difficult because recreational landings often are underreported or not surveyed. Serious allocation disputes have been limited thus far primarily

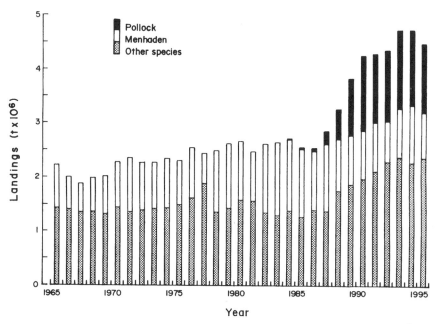

FIGURE 2-5   Total U.S. commercial landings, 1965-1995. Source: Redrawn from NMFS data.

TABLE 2-1   Comparison of U.S. Recreational and Commercial Catches for Selected Species in 1994

Fish Species	Recreational Catch (t × 1,000)	Commercial Catch (t × 1,000)
Bluefish (*Pomatomus saltatrix*)	7.2	4.4
Red snapper (*Lutjanus campechanus*)	1.3	1.5
Spotted seatrout (*Cynoscion nebulosus*)	5.1	1.1
Summer flounder (*Paralichthys dentatus*)	4.2	8.9
Winter flounder (*Pleuronectes americanus*)	0.7	3.6

to the United States and a few other industrialized nations (e.g., New Zealand), although the growth of ecotourism could create commercial-recreational fishery conflicts in industrializing nations.

Indigenous People's Fisheries

Indigenous people's fisheries are a minor part of total catches but are particularly important in cultural and social terms. Indigenous marine fisheries in the United States—primarily in Washington, Oregon, California (NRC 1996b), and Alaska—are subject to treaties between the United States and tribal groups. Tribal fisheries for salmon include commercial, ceremonial, and subsistence uses. The Northwest Indian Fisheries Commission handles treaty rights related to salmon in the Puget Sound area. The Columbia River Inter-Tribal Fisheries Commission represents four tribes in the Columbia River basin of Oregon. Fishing by three tribes in the Klamath River Basin in California is not protected by treaty, but 50 percent of Klamath River chinook salmon are allocated to these tribes by government regulation (NMFS 1996a). The Pacific Fisheries Management Council, as well as its Scientific and Technical Committee and its Salmon Technical Team, have Native American tribal representatives. There is also a Native American allocation for sablefish (*Anoplopoma fimbria*) off the coast of Washington. For communities in western Alaska, which are largely populated by Alaska natives, Section 305 of the Magnuson Fishery and Conservation Act and Section 111 of the 1996 amendments provide for a Community Development Quota Program. The program allots varying percentages of the total allowable catch (TAC) of several fisheries to these communities; by 1999 those communities will be allotted 7.5 percent of the TAC of Bering Sea groundfish and crabs. The amendments also allow the establishment of a similar program in the Western Pacific Regional Fishery Management Area (Hawaii and other U.S. Pacific islands).

Subsistence Fisheries

Subsistence fisheries—fisheries conducted for food, material, and fuel but not primarily for commerce or recreation—occur in many parts of the world, most commonly in nonindustrialized and tribal societies. Although the sale of fish for cash is not included in subsistence fishing, trading fish for other food or services is an important part of subsistence economies in many places, and cash from activities in market economies is used to finance subsistence fishing (e.g., NRC 1994a). Subsistence fishers, like others but perhaps to a greater degree, develop a large store of traditional knowledge. Subsistence fishing is recognized in many laws and regulations. It does not usually constitute a major portion of the landings except locally.

Status of U.S. Fisheries

In its most recent assessment of the condition of U.S. fisheries, NMFS (1996a) evaluated 275 stocks caught by fishers in nearshore coastal waters, the EEZ, and the high seas beyond the EEZ for the period 1992-1994. Of the 191 stocks for which information was available, 33 percent were overutilized and 49 percent were fully utilized, leaving only 18 percent underutilized. Forty-six percent were below the level of abundance required to produce the greatest long-term potential yield. The long-term potential yield of the U.S. fisheries within the U.S. EEZ is estimated on a single-species basis to be 8.1 million t per year, which is much greater than the recent yields (NMFS 1996a). Based on the calculations in that estimate, for the United States to achieve its potential increase in long-term potential yield, some ("underutilized") fisheries would need to be fished more heavily, but, more importantly, fishing on overutilized stocks, bycatch, and unaccounted mortality will need to be reduced so that stocks can rebuild. The estimated long-term potential yield and maximum sustainable-yield levels can be used as reference points to help guide the sustainable development and prosecution of fisheries or the rebuilding of marine fish stocks that have been overfished. Although there is limited information available regarding the overall economic performance of U.S. fisheries, they are undoubtedly suffering from overcapitalization at a national level (with some regional exceptions), as has been reported by FAO for fisheries worldwide. NMFS (1996b) reported that there were about 23,000 commercial fishing vessels in the United States in 1987 (the latest year for which there is good information), which is more capacity than is required to achieve the long-term potential yield from U.S. fisheries. For example, the capacity off Alaska has been estimated to be two and a half times that necessary to catch the available resources (North Pacific Fishery Management Council [NPFMC] 1992). Major losses in revenue from New England fisheries have resulted from overfishing, driven in large part by excess capacity (Edwards and Murawski 1993).

A CANADIAN EXAMPLE: NORTHERN COD

The case of the northern cod is an example of the effects of overfishing as well as institutional difficulties in applying scientific findings to management. This overfishing occurred despite reasonably conservative target fishing mortalities; the problem was largely due to systematic errors in stock assessments exacerbated by unreported (illegal) discarding of small fish and perhaps unreported catches, and later to a failure of management to respond quickly to corrected assessments.

The fisheries of the Atlantic Canada region have been dominated by groundfish (Munro 1980); the cod fishery was unquestionably of greatest importance. Cod (*Gadus morhua*) served as the base of the fishing industry in Newfoundland, Nova Scotia, and other provinces in the region (e.g., New Brunswick). The northern cod fishery is an instructive example of overexploitation of a fishery. It has been much discussed, recently by Walters and Maguire (1996) and Hutchings and Myers (1994), who focused on fishery biology, and by Neis (1992), Steele et al. (1992), and Finlayson (1994), who focused on the sociology of science. A combination of lack of data, improper handling of available data, and overconfidence in methods led to overfishing and the collapse of the fishery.

The offshore catch of northern cod expanded from approximately 240,000 t annually in the mid-1950s to a peak of 700,000 t in 1968 (Munro 1980). Total catches of northern cod declined steadily thereafter. By the early 1970s, distress in the northern cod inshore fishery also was evident. The Canadian government planned to rebuild the resource through reduced fishing by the distant water fleets. As the resource rebuilt, the total allowable catch (TAC) for northern cod would gradually be increased (Finlayson 1994). The management strategy adopted was expected to result in sustainable catches averaging 20 percent of the exploitable biomass (Canada 1990) by the late 1980s—roughly 400,000 t annually (Munro 1980). In the years immediately following implementation of this plan, the northern cod resource appeared to be rebuilding as planned, but the actual landings never achieved even 260,000 t per year. The offshore sector of the cod fishery always succeeded in taking its allocation, but the inshore sector landings declined by 35 percent from 1982 to 1986 (Buffet 1989, Finlayson 1994).

In response to these trends, the Canadian Atlantic Fisheries Scientific Advisory Committee (CAFSAC) undertook in 1988 a review of its stock-assessment methods. CAFSAC concluded that the northern cod stock was in fact substantially smaller than previously believed, a view confirmed by an independent Northern Cod Review Panel (Canada 1990), which estimated that the actual fishing mortality rates had been at least double those projected in the Canadian management strategy (Canada 1990, p. 3). CAFSAC concluded that the northern cod TAC for 1989 should be reduced from the continuing 266,000 t to 125,000 t. The Canadian government, fearful of the economic disruption and dislocation that such a draconian reduction in the TAC would entail, reduced it only to 235,000 t (Buffet 1989).

In mid-1992, after poor catches, a moratorium was established on all directed commercial fishing for northern cod for a period of two years, during which deterioration of the stock continued. The moratorium was extended, remains in place, and is expected to remain in effect for the indefinite future. Recent reports indicate that the northern cod stock is at a historically low level and that there are, as yet, no significant signs of recovery of the stock (Canada 1995).

The causes of the resource management catastrophe are the focus of intense debate in Canada (Neis 1992, Steele et al. 1992, Finlayson 1994, Walters and Maguire 1996), although the proximate cause is clearly overfishing (Hutchings and Myers 1994) supported by erroneous assessments of stock size and fishing mortality (Myers et al. 1997). The reasons for the errors in stock assessments are complex. Overcapitalization in the fishery may have exerted pressure to interpret stock-assessment data in an excessively optimistic manner (Canada 1990), as did an overreliance on the science and culture of quantitative stock assessment (Walters and Maguire 1996, Finlayson 1994). The northern cod stock-assessment procedures appear to have been flawed from 1977 until at least 1985, owing to statistical inadequacies of the biomass model used, overreliance on catch-per-unit-effort data, variability of the data set, and relatively short and unreliable data series (Canada 1990, Walters and Maguire 1996). Ironically, the northern cod is one of the few examples that seems to show a clear and positive relationship between parent stock and recruitment (Hutchings and Myers 1994). Although the intuitive expectation is that the more spawning adults there are in the population, the more recruits there will be, most fish populations do not show such a relationship. For the northern cod, estimates of recruitment were not corrected for changes in spawner biomass, which themselves were overestimated (Hutchings and Myers 1994). Fishing mortalities were underestimated, probably because of unreported discards of young fish—a significant source of mortality—and perhaps unreported or underreported catches of adult fish (Myers et al. 1997).

Recent analyses indicate that the cod populations in the western Atlantic are not the only ones in danger of being overfished. For example, Cook et al. (1997) concluded that there is an urgent need to reduce the exploitation rate on North Sea cod to avoid risk of collapse (they also found a significant relationship between parent stock and recruitment). Several NMFS assessments of cod in the Gulf of Maine also reached this conclusion, recently confirmed by the National Research Council (NRC 1998b).

CONCLUSIONS

Global marine fish catch is at or near its sustainable limit. Many species and some regions are seriously overexploited. Populations of long-lived, slow-growing species are especially vulnerable to collapse as a result of overfishing. The estimates are primarily based on single-species considerations; as described in later chapters, consideration of fishing's effects on biological communities and ecosystems and the need to balance a variety of societal goals reinforces the conclusion that a sustainable general increase in the yield of marine fisheries is probably not possible. Indeed, a moderate level of exploitation may be a better goal for fisheries than full exploitation, because full exploitation tends to lead to overexploitation. Under this strategy relatively few fisheries worldwide (i.e., the relatively few commercial stocks that are lightly fished) are good candidates for increased exploitation. Better management is possible, however, and could greatly improve the situation.

3

Fishing and Marine Ecosystems

INTRODUCTION

Ecosystems are complex, linked, interactive systems in which organisms, habitats, and external forcing (e.g., ocean currents, weather) act together to shape communities and regulate population abundances. Fishing is a major activity that can selectively remove large portions of animal populations and also significantly alter trophic interactions. Fishing gear that drags the bottom at times, such as bottom-trawls, pots, and longlines, can alter marine habitats, especially benthic or reef habitats. Because fish populations fluctuate naturally, sometimes by orders of magnitude, separating the effects of fishing from natural biological and environmental variability is difficult.

In analyzing the effects of fishing and environmental factors on populations and ecosystems, the committee did not try to be exhaustive. Instead, enough information is provided to be representative. One difficulty is that fishing has been pursued for many years in many marine ecosystems, so that most "baselines" represent ecosystems that already are much affected (Pauly 1995, Jackson 1997). Controls are difficult to establish, and without them, the effects of fishing can be difficult to gauge (Roberts 1997).

Fishing alters the age and size structure of populations. A consequence of fishing is an altered demography. Older and larger fish often are removed first, and the remaining older cohorts experience more cumulative fishing (Baranov 1918), and the population becomes more and more dependent on small, newly recruited individuals to support the fishery. In many ecosystems, large piscivorous species were the initial targets of fishing. After the "fishing down" of those populations, the fisheries shifted to smaller species at lower trophic levels (Pauly et al. 1998). The net effect is a major change in community structure through altered trophic interactions. Not only do fish abundances and biomasses decline, but the entire ecosystem structure can be changed. Fishing also can alter the genetic structure of fished populations. Although the immediate results of such genetic alteration might be hard to detect (Policansky 1993a), the large-scale changes that can occur in widespread fish species, especially those with discrete subpopulations like Pacific salmon, can substantially reduce the species' ability to recover from the effects of depletion due to fishing and other causes (Policansky and Magnuson 1998, NRC 1996b).

To understand the effects of fishing in an ecosystem context, the ecology of removed fish must be examined. Not surprisingly, the majority (81 percent by weight) of the 50 major species in marine catches (excluding cultured organisms) (FAO 1996a) are the small and abundant animals that are low on the trophic pyramid, eating phytoplankton (40 percent), zooplankton (38 percent), and algae (3 percent). These planktivores may have an important role in controlling plankton productivity and community structure. For example, it is hypothesized that overfishing, habitat changes, and changes in water quality associated with the removal of oysters from Chesapeake Bay (Rothschild et al. 1994, Lenihan and Peterson 1998) caused a cascade of events that may have promoted explosions of ctenophores (comb jellies) (Newell 1988). The decline of oysters may have contributed to increased phytoplankton biomass and concomitant decreases in sunlight penetration that led to declines of benthic seagrasses, which were important nursery areas for many species. Ulanowicz and Tuttle (1992) developed a network model which indicated that re-establishment of oysters and control of fishing mortality on them would increase benthic primary production, fish abundances, and zooplankton densities. In addition, oyster reefs provide physical structure and habitat for a variety of invertebrates and fish. Another example, that of removing herbivores from coral reefs, is discussed below.

Although most marine landings are of small planktivores, fishing and the removal of large species at high trophic levels that affect ecosystem structure and functioning (i.e., piscivores and large predators) can have major effects on ecosystems (sometimes called "top-down control") (Carpenter et al. 1985, Hixon and Carr 1997). It has been hypothesized and demonstrated in some systems that removal of predators can have impacts out of proportion to their abundances through a trophic cascade that affects not only prey of predators but also indirectly the lower trophic levels that constitute the food resource of the small fish and invertebrates that were eaten by predators (e.g., Paine 1980, Simenstad et al. 1978, Carpenter and Kitchell 1988, Parsons 1992). Marine assemblages dominated by sea otters (Box 3-1) are useful model systems for understanding and describing multispecies, multitrophic-level interactions. Three trophic levels— carnivorous otters, herbivorous urchins, and photosynthetic benthic algae—are clearly present. Because otters can control the density and size structure of the urchins and the urchins control the specific identity and therefore the productivity of the algae, the linkage between the species and trophic levels is strong. At high density, the otters instigate a trophic cascade (Paine 1980).

Trophic cascades are characterized by three features generalizable to other benthic and aquatic systems: high trophic-level species influence the assemblages' structure (top-down control), indirect effects two or more links distant from the primary one are biologically conspicuous, and alternative community states in which different species are abundant and ecologically dominant (Sutherland 1974, Hughes 1994) can persist. To the degree that these results from benthic assemblages can provide an analogy for the dynamic organization of nearshore and oceanic communities, they provide a cautionary note. Fishery biologists should anticipate community changes when high trophic-level populations are heavily exploited. This effect is not necessarily only a consequence of modern technology. For example, Aleut exploitation of sea otters appears to have changed the structure of nearshore marine communities as long 2,500 years ago (Simenstad et al. 1978).

In addition to the effects of directly removing animals and the effects on the ecosystem that this precipitates, fishing can physically affect the marine environment. The most prominent of such effects are destruction and disturbance of bottom communities and of bottom topography by trawls and dredges (Dayton et al. 1995, Auster et al. 1996). The advent of large-scale mariculture and the development of offshore facilities to support it also have the potential to alter marine ecosystems significantly.

Both long- and short-term environmental fluctuations have major effects on the abundances of marine organisms (e.g., Soutar and Isaacs 1974, Baumgartner et al. 1992). Although not caused by fishing, environmental effects are often hard to distinguish from those caused by fishing (e.g., Wooster 1983; Bakun 1993; Cushing 1982; Rothschild 1986, 1995) and can have profound consequences for management. For this reason, the following sections include a brief review of environmental effects as well as more detailed reviews of the effects of fishing.

REMOVAL OF HERBIVOROUS FISHES FROM CORAL REEF ECOSYSTEMS

One of the more dramatic ecological effects of removing herbivores, including fishes, from an ecosystem has been described by Hughes (1994), Jackson (1997), and others. Caribbean coral-reef ecosystems include herbivorous fishes and urchins (*Diadema antillarum*) as well as several species of coral. However, many of the herbivorous fishes and other animals had been removed by fishing long before any serious ecological study of coral reefs began in the 1960s, so the importance of their presence was not appreciated until a combination of recent events made matters clearer. Indeed, Jackson (1997) considered the study of reef herbivores since the 1950s to be analogous to studying termites and locusts in the Serengeti in the absence of wildebeest and elephants.

The first event in the chain that exposed the effects of chronic overfishing was Hurricane Allen, a category 5 hurricane that struck the Caribbean in 1980. It

BOX 3-1
Sea Otters

Northern sea otters (*Enhydra lutris*) originally ranged across the Pacific rim from Hokkaido (Japan) to Baja California (Mexico). Their primary prey includes a broad range of invertebrates and fishes; because they inhabit nearshore environments, a wealth of ecological detail is available (Van Blaricom and Estes 1988). Exploitation for their pelts led to near extinction by 1911, when unregulated killing ceased. Preservation and more recently, restoration, rather than population sustainability have become the dominant themes of otter biology. During early phases of recovery, otter populations grew at a rate of about 15 percent per year, especially in the Aleutian Island chain, producing high population density on some islands in the same geographic region as islands that lack otters. Inter-island comparisons were central to the classic study of Estes and Palmisano (1974): otters as keystone species control the local biomass and, to a lesser degree, the abundance of sea urchins, which regulate benthic algal biomass and productivity. Duggins and colleagues have increased our confidence in these patterns of direct interactions by observations made before and after reinvasions demonstrating that in otter-dominated habitats, high kelp biomass and therefore the generation of detrital materials have also indirectly increased the growth rate of suspension-feeding invertebrates (Duggins 1980, Duggins et al. 1989). Otter populations increased generally until about 1990 throughout western Alaska and the Aleutian Islands; since then a decline amounting to 35 to 90 percent of the populations has occurred. Although the causes remain unknown, the consequences increase our confidence in the robustness of the classic otter-induced cascade. At Adak Island, where the otter population has declined from an estimated 5,000 to slightly more than 600 in 1997, urchins have increased about fivefold in number, and their maximum size has more than doubled. Conversely, kelp density has declined by a factor of 10 (James Estes, University of California, personal communication, 1997).

It seems unlikely that aboriginal Aleut hunters maintained sustainable otter populations, based on cycles of apparent abundances and associated community changes identified in midden remains (Simenstad et al. 1978). The otter provides a textbook study of the (unresolved) difficulties that ecosystem managers may face, because legally mandated protection of otters conflicts with multiple-use concepts (Levin 1988). Otters consume and are capable of local decimation of sea-urchin populations (the exploitation of which is a growing multimillion-dollar industry), although in the process the otters indirectly facilitate kelp production (itself an important commercial resource). Kelp beds provide nursery grounds for nearshore fishes, and their detrital production may enhance the growth of abalone or other benthic invertebrates (e.g., Pismo clams). Otters consume these animal species, many of which are of recreational or commercial significance and clearly play a critical role in the structuring and organization of nearshore marine ecosystems.

severely damaged coral reefs around Jamaica that had not seen a major hurricane for more than 40 years. For the next three years the corals recovered, but in 1983 a disease devastated the *Diadema* populations. Because both the major groups of herbivores were now absent—the fishes, reptiles, and mammals removed by exploitation and the urchins killed by disease—the recovery of corals stopped because they became overgrown by algae. The coral cover was reduced from a mean of 52 percent in 1977 to 3 percent in the early 1990s, and the cover of fleshy macroalgae increased from 4 to 92 percent. Although the events described by Hughes (1994) occurred in Jamaica, other areas of the Caribbean have been similarly affected, and much of the Caribbean has been subjected to overexploitation of herbivores for at least several hundred years (Jackson 1997). High nutrient loads that favor algal growth over coral growth also have adversely affected the region's marine ecosystems.

Fishing has devastated coral-reef ecosystems in the Pacific as well. Although illegal, dynamite is often used there. Johannes and Riepen (1995) reported on fishing with cyanide to catch live reef fishes in the Philippines, Malaysia, and Indonesia. Cyanide fishing is inefficient because many fish die in reef crevices and are not captured; of those captured, some die before reaching their destinations and are not used. The fishing technique also kills the reefs and nontarget species, so its effect on reef ecosystems is even greater than the loss of the fish.

Other Ecosystem Effects of Fishing in the Philippines

The Philippines were the first among Southeast Asian countries to develop a modern bottom-trawl fishery immediately after World War II. This fishery grew quickly, as did the pelagic fishery, which relied mainly on boats using light sources to attract fish into nets, and a coral reef fishery using a variety of mostly destructive methods, notably the persistent *muro-ami* technique of pounding reefs to scare fish into surrounding gillnets.

The Philippines consist of about 7,000 mostly high islands, surrounded by a relatively small continental-shelf area. The adjacent waters are extremely deep and infertile. The inherent limits to fishery catches implied by the small shelf area were until recently not recognized by government planners and development banks, which subsidized acquisition by the industrial sector of a fleet capacity about three times in excess of what was required to harvest the present marine catches of 1.9 million metric tons (t) per year (BFAR 1994).

The excess capacity in Philippine fisheries is particularly tragic in that it impoverishes the numerous small-scale fishers, who would be technically capable of catching and marketing most of the fish presently harvested by the overcapitalized industrial fleet. This impoverishment, further exacerbated by large numbers of landless farmers entering the small-scale fishery sector, leads to what is now called "Malthusian" overfishing (Pauly et al. 1998), characterized by the widespread use of dynamite and cyanide as cheap ("entry-level") fishing gears. The effects on coastal ecosystems of the combined fishing pressure of the industrial and small-scale sectors, fishing techniques that destroy reefs, and the downstream effects of nonsustainable agricultural and forestry practices have been devastating, with most of the fringing reefs surrounding the Philippine islands choking under silt and experiencing massive species changes. The slowly declining fish supply is now dominated by small, low-value species.

BYCATCH, DISCARDS, AND UNOBSERVED FISHING MORTALITY

Bycatch is the capture of nontarget species in directed fisheries. Discards are animals returned to the sea after being caught. Some bycatch is retained in most fisheries, but most of it is usually discarded and not reported in official landing statistics. Even some retained bycatch is not reported, and sometimes targeted species are discarded if they are too small. It is important to be clear about these terms (Alverson et al. 1994) because they are often used differently by different authors, and it is impossible to evaluate the importance of bycatch and discards if they are not clearly distinguished from each other.

Because a significant portion of the catch (discards) in many fisheries is not reported, the portion that *is* reported constitutes only a fraction of the actual catch. Discards make stock assessments and fishing mortalities difficult to estimate because they are usually unreported; even when they are, the reports usually lack specific information on the age and size of the animals discarded, which is needed for stock assessment. Thus, inferred patterns of exploitation are significantly underestimated. The situation is made worse by a variety of other unobserved sources of fishing mortality such as illegal fishing, underreporting, deaths of fish that escape from fishing gear, and ghost fishing. Recreational and subsistence fishing are also difficult to monitor and thus can represent a significant source of unobserved fishing mortality, especially for widely sought species such as bluefish (*Pomatomus saltatrix*), spotted seatrout (*Cynoscion nebulosus*), and summer flounder (*Paralichthys dentatus*) in the southern and eastern United States. The magnitude of discard mortality and unobserved fishing mortality could be important factors contributing to global overfishing and undesirable ecological changes.

Alverson et al. (1994) reviewed the literature on worldwide bycatch and discards and concluded that marine discards in the period 1988-1990 amounted to approximately 27 million t per year, roughly one-third as much as the total marine capture fisheries.[1] There is great variation in bycatch associated with

---

[1]Bycatch and discards are extremely difficult to estimate precisely, in part because they often are illegal or unregulated activities. The estimates of bycatch by Alverson et al. (1994) had a low limit of 17.9 million t and a high limit of 39.5 million t; more recent estimates suggest that bycatch has decreased (D.L. Alverson, personal communication, May 1998).

various kinds of fisheries. By far the largest contributor to marine discards—nearly 10 million t per year—was shrimp trawling, especially in tropical regions. This biomass of discards represents approximately five times the biomass of the shrimp and includes more than 240 species, including the young of commercial species and adults of some species that mature at less than 10 cm in length (Alverson et al. 1994).

On the other hand, not all trawl fisheries have high bycatch rates. In the trawl fishery for northwest Atlantic silver hake (*Merluccius bilinearis*), bycatch is about 1 percent of landings by weight (Alverson et al. 1994). In some cases the very large catches of target species can result in large absolute bycatches and discards even when the bycatch rate is low. One example is the midwater-trawl fishery for walleye pollock (*Theragra chalcogramma*) in the Bering Sea and the Gulf of Alaska, where overall discards are reported to be approximately 6 percent of the landed weight and 0.5 percent of the landed number of animals, but discards in 1992 involved more than 130 species and exceeded 100 million animals, more than half of which were pollock. Aggregate discards in the Bering Sea and Gulf of Alaska bottom fisheries approach 1 billion animals each year (Alverson et al. 1994). Another example is the Gulf of Mexico shrimp-trawl fishery, which has resulted in the deaths of tens of thousands of sea turtles, but the bycatch rate is so low that many shrimpers do not catch even one turtle in the course of a year (NRC 1990).

In some fisheries the species in the bycatch are of special concern because they contribute to and aggravate an overfishing problem, involve the target species of other highly regulated fisheries (e.g., Pacific halibut [*Hippoglossus stenolepis*] caught in bottom-trawl or crab-pot fisheries in the region), or are threatened or endangered species (e.g., turtles caught in shrimp trawls). Bycatch mortality and the resulting discards can have a significant effect on a particular nontarget species or on a marine community. For example, the red snapper (*Lutjanus campechanus*) taken in the Gulf of Mexico shrimp fishery presents a convincing example of how bycatch losses can affect a valuable nontarget species. Young red snapper suffer heavy bycatch mortality in the trawl fishery for shrimp (*Penaeus* spp.) (Alverson et al. 1994). The discarded bycatch represents the single largest component of fishing mortality on the red snapper population.

Using present trawling gear and methods in combination with bycatch-reduction devices, prohibiting trawling in some places at some times, and limiting towing times would probably decrease landings and revenues from the shrimp fishery but increase landings and revenue to red snapper fishers. The societal tradeoffs involved are not clear, but studies suggest that if a substantial reduction in bycatch were achieved, a sustainable commercial red snapper fishery with landings at least three times higher than those now recorded would be possible (Goodyear 1985, 1995; Goodyear and Phares 1990). A 50 percent reduction in red snapper bycatch has been recommended by the Gulf of Mexico Regional

Fishery Management Council, but no regulatory action has yet been taken to achieve that goal.

Other well-known examples of bycatch problems have been or are being addressed by technological and operational changes in fishing. One example is the killing of dolphins in the yellowfin tuna fishery in the eastern tropical Pacific (NRC 1992a, MMC 1998). Formerly responsible for hundreds of thousands of dolphin deaths per year, the fishery now kills fewer than 3,000 dolphins annually.[2] Another example is shrimp trawling, which was killing as many as 11,000 endangered sea turtles a year in the Gulf of Mexico and off the southeastern U.S. coast (NRC 1990), but the adoption of turtle-excluder devices has reduced that mortality. Many examples of bycatch and potential solutions to them were discussed at a recent workshop (Alaska Sea Grant 1996).

The ecological consequences of bycatch and discards are not well quantified for most marine ecosystems. As described in Chapter 5, more research is needed on those consequences. Nonetheless, there appears to be general agreement that bycatch and discards can and should be reduced (Alverson et al. 1994, Alaska Sea Grant 1996). Chapter 5 describes various current efforts to reduce them as well as suggestions for additional approaches.

FISHING AND LARGE MARINE ECOSYSTEMS

Georges Bank

Georges Bank—a shallow, productive, submarine plateau off New England and Nova Scotia—is the poster-child, so to speak, for the effects of overfishing. In addition to severe depletion of several populations of commercially valuable groundfish, the Georges Bank ecosystem and bottom habitat have suffered large impacts from fishing. The following account is adapted from Fogarty and Murawski (1998).

Georges Bank has supported commercial fishing for at least four centuries. Important finfish species included halibut and other flatfish species, especially yellowtail flounder (*Pleuronectes ferrugineus*), haddock, and cod. The two major impacts on the fishery were the arrival of the distant-water fleet in 1961 and the modernization and expansion of the domestic fleet after the establishment of extended U.S. jurisdiction over fisheries (the 200-mile limit) in 1977. Despite repeated warnings from scientists, the principal groundfish species have been severely depleted; Fogarty and Murawski (1998) reported that exploitation rates of cod and yellowtail flounder were 55 and 65 percent, respectively, of the

---

[2]This number of dolphins killed is far below the number that their populations can sustain (NRC 1992a), although U.S. policy as expressed in the Marine Mammal Protection Act (P.L. 92-522 as amended) and the recently passed International Dolphin Conservation Program Act (P.L. 105-42) has an ultimate goal of zero dolphin deaths.

biomass in the early and mid-1990s;[3] the recommended exploitation levels based on optimal sustained yield were 13 and 22 percent, respectively. Spawning-stock biomasses were also very low, about 10 percent or less of their values in the 1950s for haddock, cod, and yellowtail flounder.

The depleted groundfish populations have largely been replaced by populations of small elasmobranchs, mainly spiny dogfish (*Squalus acanthias*), smooth dogfish (*Mustelis canis*), and skates (*Raja* spp.). Other changes in the pelagic fishes included initial declines of mackerel and herring and concomitant increases in American sand lance (*Ammodytes americanus*), their prey. Recently, fishing pressure has resulted in declines in the elasmobranch populations—the species are mostly long lived and slow growing compared to many commercially important teleosts—and mackerel and herring populations have increased as well. This has been accompanied by a decline in sand lance populations.

This example makes the ecosystem consequences of fishing clear: the groundfish community first became dominated by small sharks (e.g., dogfish) and rays, and then overfishing reduced those populations. It is not yet known whether this ecosystem will recover to its preexploitation structure in the absence of fishing or whether it will attain some other composition.

The Bering Sea

Recent declines of many populations of marine mammals and birds that live in and near the Bering Sea, a semienclosed basin of the North Pacific Ocean between Alaska and Russia, have attracted attention and have been attributed by many to the effects of fishing. The National Research Council recently reviewed the information (NRC 1996a) and concluded that fishing probably has affected the ecosystem but in a more complicated fashion than simple overfishing and in combination with environmental changes. Documented changes include changes in abundances of many fish species and changes in the physical environment. There also is persuasive (although not conclusive) evidence that marine mammals and birds are declining because the juveniles are short of food.

The NRC report concluded that the changes in the Bering Sea ecosystem were probably caused by a combination of changes in the physical environment coupled with heavy exploitation of components of the system (whales and fishes). Many sperm and baleen whales were removed from the Bering Sea and adjacent waters in the 1950s, 1960s, and 1970s. Various flatfish species, Pacific Ocean perch, and herring were also heavily fished in that period, with resulting population declines. Many of those species feed heavily on zooplankton and thus compete with walleye pollock; others of those species prey on pollock. In the late

---

[3]The exploitation rates are much lower following recent action by the New England Fishery Management Council (NEFMC; see NRC 1998b and the Northeast Multispecies Fishery Management Plan and amendments at NEFMC's web site, www1.shore.net/~nefmc/.

1970s the physical regime appears to have shifted as well, resulting in higher sea-surface temperatures and less ice cover than before, conditions that seem to favor pollock recruitment.

As a result, the ecosystem appears to have become more dominated by pollock than it was before. In recent years other predatory fishes—mostly flat-fishes—have increased as well. These predators might be responsible for the decline of species normally favored by marine mammals and birds, such as capelin (*Mallotus villosus*), Pacific sand lance (*Ammodytes hexapterus*), and squid (*Berryteuthis* sp. and *Gonatus* sp.) As a result of these changes, juvenile marine mammals and birds have been deprived of their preferred foods. Thus, fishing (including whaling) appears to have contributed to a significant change in the structure and functioning of this large marine ecosystem, although the total biomass removed by the current pollock fishery does not seem to be a major contributor to the problem.

Analogs to the Bering Sea

The complexity of marine ecosystems and the number of potential factors involved make it difficult to have great confidence in our understanding of the precise mechanisms that relate fishing to the populations of top predators. The NRC report (NRC 1996a) pointed out that, although there has been heavy fishing pressure in the North Sea, in the upwelling areas off South Africa and Namibia, and off Peru, there have not been clear effects on the populations of pinnipeds.

Nonetheless, it seems likely that continued removal of large portions of various trophic levels from marine ecosystems will affect ecosystem structure and functioning. One issue that needs resolution is the effects on marine ecosystems of populations of marine mammals as they recover from very heavy exploitation. Baleen whales eat zooplankton and thus compete for food with many commercially important fish species; toothed whales eat fish and squid and thus compete directly with humans for food. The recovery of whale populations is one of many examples where different policy goals (i.e., protection of marine mammals, catching fish for food) have the potential to conflict.

The Barents Sea

The Barents Sea, off the extreme northwestern coast of Russia and the extreme northern part of Norway, contains heavily exploited populations of cod (*Gadus morhua*), capelin (*Mallotus villosus*), and herring (*Clupea harengus*) as well as marine mammals (whales and seals). Collapses of fish populations, crises in the fisheries, and a destabilization of the ecosystem occurred during the 1980s, a consequence of overfishing in the Norwegian and Barents seas.

The mature stock of Atlanto-Scandian herring is fished primarily in the Norwegian Sea, but its young use coastal regions of the Barents Sea as a nursery, where herring is an important prey of cod. Capelin is an opportunist species that is also an important prey of cod. All of the species experience good recruitment when temperatures are above average, which occurs during periodic events when substantial Atlantic water flows into the Barents Sea (Skjoldal et al., 1992; Hamre, 1994).

Atlanto-Scandian herring were overfished in the Norwegian Sea and areas outside the Barents Sea. The herring stock collapsed, and its recruitment failed during the 1960s and 1970s. The scarcity of herring, itself a major predator on larval and small juvenile capelin, favored recruitment and growth of the capelin population. Despite heavy fishing, the capelin population declined only slowly during the 1970s when temperatures were below normal. Capelin grew slowly and matured at relatively old ages, minimizing massive postspawning dieoffs. The cod population, which sustained continuous heavy fishing, declined from the mid-1970s through the early 1980s.

Paradoxically, Atlantic inflow events in the early 1980s triggered the ecological crisis. Herring pre-recruits (fish still too small to be targets of fishing) returned in modest numbers to the Barents Sea, not in high enough abundance to satisfy the predation demand of cod but numerous enough to cause recruitment failures of capelin during the mid-1980s. Higher temperatures and fast growth of previously recruited capelin during this period brought them to maturity; they spawned and died. As a result of this loss of mature capelin and the herring-induced recruitment failure, the capelin population collapsed. The strong year classes of cod in the warmer environment had insufficient capelin or herring prey to satisfy demands for growth; their condition deteriorated, cannibalism became common, and the biomass of cod declined (Laevastu 1993). The fisheries for cod and capelin were in crisis.

The capelin fishery was placed under moratorium from 1987 to 1990. The opportunist capelin made a strong recovery in the early 1990s, perhaps associated with warming of the ocean (Everett et al. 1996), but the populations collapsed again in the mid-1990s (Institute of Marine Research, Bergen 1995). The Barents Sea ecosystem remains in stress.

The relatively simple community structure of the Barents Sea ecosystem is easily destabilized. Even in "simple" systems, interactions among species and the role of environment are complex and initially unpredictable. Herring, which apparently play a pivotal role in controlling Barents Sea dynamics, were over-fished in the adjacent Norwegian Sea. That spatially removed intervention, combined with complex changes in ocean climate and plankton dynamics (Skjoldal et al., 1992), set the stage for the collapse of Barents Sea fisheries.

The Southern Ocean

The Southern Ocean—the seas off Antarctica south of approximately 50 to 60 degrees—is substantially isolated from other oceans by the Antarctic Convergence or South Polar Front (Kock 1992). It is noted for the presence of the world's heaviest seabirds (penguins), large populations of krill (planktonic crustaceans, mainly *Euphausia superba*), formerly large populations of whales, and some commercially valuable fish species. Most of the fishes are relatively slow growing and thus cannot support high exploitation rates (Kock 1992). Indeed, Kock estimated that the total sustainable catch of all finfish from the Southern Ocean is no more than 100,000 t per year, about 0.1 percent of the world catch. As mentioned in Chapter 2, the marbled rockcod (*Notothenia rossi*) has been severely overexploited, with 400,000 t being taken in 1969-1970 alone. Recent production has been on the order of a few thousand metric tons per year. The Patagonian toothfish (*Dissostichus eleginoides*) also appears vulnerable to overexploitation, which has probably already occurred. It has become popular in restaurants, where it is known as Chilean sea bass. Catch rates and population estimates of this fish are poorly known (Kock 1992, Albemus 1997), but Kock reported that the some *D. eleginoides* populations were significantly overfished.

Although ecosystem effects of overexploitation have probably occurred in the Southern Ocean, they have unfortunately not been well studied (Kock 1992). Despite the general lack of information, it is known that the main prey species for larger marine vertebrates in the ecosystem are krill, and they have been studied extensively (Miller and Hampton 1989). They appear to be larger, longer lived, and slower growing than most marine plankton. Biomass estimates vary, but the standing stock is probably at least several hundred million metric tons in the summer (Miller and Hampton 1989). Calculations by Bengston (1984), Laws (1985), and Yamanaka (1983) and a detailed review by Miller and Hampton (1989) suggest that current consumption of krill by predators in the Southern Ocean exceeds 200 million t per year, of which perhaps 40 million t is due to baleen whales. Baleen whale populations in the Southern Ocean consumed perhaps 190 million t per year before exploitation. Thus, there is a "krill surplus" of about 150 million t per year (Miller and Hampton 1989). This is not currently being exploited by fisheries (krill fisheries are only around 100,000 t/year [David Agnew, Imperial College, London, personal communication, 1997] and finfish landings are very low, as mentioned above). It would be of great interest to know how the "surplus" is being consumed in the ecosystem and how recovery of whale populations would affect the ecosystem. As Miller and Hampton (1989) pointed out, it is not really a surplus; the term refers only to krill that is no longer consumed by whales. Some of the "surplus" may be recycled through decomposition pathways; most of it probably is eaten by other predators. Thus, the removal of whales has led to increases in other predator populations. It would also be of interest to know how recovery of whale populations in other areas with significant fisheries, such as the Bering Sea, might affect the ecosystem.

Several features of the Southern Ocean ecosystem make it a good system in which to study the ecological effects of fishing. The ecosystem has short food chains and thus a great abundance of top predators, such as seabirds, whales, and seals (Center for Law and Social Policy and The Oceanic Society 1980). It has received a great deal of attention and has experienced relatively little exploitation to date (with the obvious exception of the great whale fisheries). Drawbacks include its large size and lack of detailed fishery statistics.

DEEP-SEA FISHERIES

The deep ocean (more than 1,000 m) has an enormous volume, comprising more than 88 percent of the world's ocean (Merrett and Haedrich 1997). It contains many species of fishes and invertebrates, some of which have large populations (Merrett and Haedrich 1997, Haedrich 1997). As technology and fishing power have improved and as some populations of commercially important species have become less abundant in shallower waters, pressures have increased to fish deeper waters. Because most midwater fishes that live at great depths are small, much attention has been given to the larger demersal species, the best known of which is the orange roughy (*Hoplostethus atlanticus*) (Merrett and Haedrich 1997).

Merrett and Haedrich (1997) took an ecosystem approach to evaluating the potential of the deep sea to provide a significant yield of fish. They acknowledged that information on many aspects of deep-sea fishes and fisheries is sparse, but they concluded that the available information suggests that no large-scale fishery for deep-sea fishes is possible. Several factors support this conclusion. First, the energy available to deep-sea ecosystems is much less than that available for shallower marine ecosystems. Second, many deep-sea fishes have watery muscles, and as much as 90 percent of their biomass is water. As a result, although they can be quite large, many deep-sea fishes yield very little edible protein. Third, and perhaps most important, the growth rates of most deep-sea fishes are low, so they become big enough to catch profitably only at ages of 30 years or more. This means that only very low exploitation rates are sustainable; indeed, Merrett and Haedrich provided many examples of deep-sea fisheries that have substantially and rapidly reduced the biomass of the target species.

Finally, Merrett and Haedrich examined the community structure of various marine ecosystems and noted that many shallow ecosystems have several top predators other than human fishers—for example, marine mammals, large sharks, and large birds. In such ecosystems, humans are competitors with those predators. Indeed, their activities seem to affect the abundance of those predators (e.g., NRC 1996a). But in deep-sea ecosystems, such predators do not seem to be present in any abundance. In Merrett and Haedrich's words (1997 p. 226), "the implication is that the addition of a higher predator, i.e., the fishery, may not be possible."

The above conclusion does not imply that no sustainable deep-sea fisheries are possible. A few localized fisheries have been conducted for many decades, apparently sustainably, such as the Madeira fishery for black scabbardfish (*Aphanopus carbo*) (Merrett and Haedrich 1997). But technological improvements and economic considerations such as the need to recoup capital investments have generally led recent large-scale deep-sea fisheries to deplete their target species. Management of such fisheries is made especially difficult by the lack of information, the lack of institutional authority when the fish are outside all countries' jurisdictions, and the very long generation times of most deep-sea fishes. Those factors make it even harder than usual to evaluate the effects and effectiveness of management.

EFFECTS OF FISHING ON BENTHIC ECOSYSTEMS

In addition to catching fish and other target species, fishing can affect ecosystems in other ways, most significantly through mechanical changes to the bottom caused by dragging fishing gear across it. Auster et al. (1996) described the effects of mobile bottom-fishing gear on the seafloor communities of the Gulf of Maine. At Swans Island, a conservation area that had been closed since 1983 was compared in 1993 with adjacent sites outside the area. At Jeffreys Bank, surveys made in 1987 were compared with surveys in 1993, after a new type of mobile bottom-fishing gear had allowed fishing on the rock-strewn bottom. Stellwagen Bank, a heavily fished area, was observed in 1993 and 1994. At all sites fished, significant and large reductions in various components of the benthic community had occurred. In some cases the changes (and losses of animals) were observed in the paths of scallop dredges and trawls; at Jeffreys Bank, the previously protected areas (because of the large rocks) showed large losses of benthic communities after the new gear allowed fishing there.

Auster et al. (1996) and Dayton et al. (1995) reviewed literature showing significant effects of bottom-fishing gear, including reduction of habitat complexity and destruction of physical refuges for animals (including biotic structures such as the tubes of tube-worms). They concluded that those processes directly reduced the diversity and productivity of many benthic communities and indirectly could affect such processes as recruitment, growth, and reproduction of many species, including commercially important ones. It seems likely that large areas of continental-shelf waters worldwide have been affected by bottom fishing, and some deepwater areas (up to 1,000 m) could have been affected as well, depending on the gear rigging and type and the substrate. Lenihan and Peterson (1998) described the degradation of oyster reefs caused by oyster-dredging in Chesapeake Bay and North Carolina's Neuse River, and showed that the physical degradation interacts with water quality to cause the observed pattern of oyster mortality. They concluded that interactions among environmental disturbances imply a need for integrative ecosystem-based approaches to restoration of estuaries, and the following sections emphasize that their conclusion applies to coastal marine areas in general.

MARICULTURE

Mariculture, or marine aquaculture, is the farming of fish, molluscs, crustaceans, and plants. It has a very long history but has grown rapidly in recent decades (NRC 1992b, FAO 1997a, 1997c, Anderson 1997, Chamberlain 1997, Lu 1997). It has often been controversial, and many U.S. state and federal laws express various policies toward it, some cautionary (NRC 1992b). We consider it here only briefly. The major and closely related issues relevant to this study are the potential of mariculture to supplement marine capture fisheries and its environmental effects.

Mariculture (excluding plants) produced 6.7 million t in 1995, about 7 percent of the world's marine fish production (FAO 1997a). (Freshwater aquaculture, also excluding plants, produced 14.6 million t in 1995.) In some cases, however, mariculture comes closer to dominating: by 1994, farm-raised shrimp production totaled 891,000 t, almost half the world's production of 1.9 million t of wild-caught shrimp (Chamberlain 1997). Farmed salmon production exceeded 550,000 t in 1995 (Anderson 1997). By 1994 it exceeded U.S. (including Alaska) commercial landings for the first time (Johnson 1995). Cultured organisms usually are higher-valued species than the average of wild-caught species (FAO 1997a), so their total value is a greater fraction of the total value of world fisheries than would be suggested by comparing total biomasses. What does the future hold? The NRC (1992b) suggested that if mariculture continued to grow at the same rate as in the recent past, it would produce 33 million t by the year 2000 and could effectively supplement a commercial fishery yield of 100 million t. Others also have given high priority to expanding mariculture (e.g., FAO 1995b).

Many difficulties that accompany the promotion and growth of mariculture have been described (e.g., NRC 1992b, Anderson 1997, Chamberlain 1997), including environmental pollution, spread of disease, introduction of unwanted exotic species, and social and economic factors. A major environmental problem is the use of coastal habitat for shrimp farming, mainly in Asian countries, which results in the loss of native coastal ecosystems, especially mangroves (Chamberlain 1997).

One important question not addressed until recently is the degree to which mariculture depends on ecosystem subsidies from elsewhere (Folke et al. 1998). In other words, mariculture—with its production of wastes that must be dissipated by local ecosystems, and its demand for food that must be created by primary and secondary production in more distant, usually marine, ecosystems (Folke and Kautsky 1992)—does not necessarily represent a more efficient use of ecosystem services than marine capture fisheries. To the degree and in the circumstances where it does, it does have potential for supplementing marine capture fishing. It is certain that terrestrial agriculture is more efficient at producing carbohydrates than natural terrestrial production, but even in terrestrial agriculture, local natural production of carbohydrate through photosynthesis is heavily subsidized by the use of fossil fuel to produce fertilizers, pesticides, and power machinery. That subsidy represents production of distant ecosystems, in this case distant in time. We do not have all the information at present to calculate and compare productivity of mariculture with marine capture fisheries, but Folke and Kautsky (1992) and Folke et al. (1998) provided some insights.

Folke and colleagues estimated that fish farming in cages uses the production of marine ecosystem areas between 20,000 and 50,000 times as large as the cages for food and areas 100 to 200 times as large for the dissipation of wastes. Folke and Kautsky (1989) calculated that intensive systems are much more demanding of ecosystem services than extensive ones and require approximately the same amount of production from natural ecosystems as capture fisheries. For example, mussel long-line rearing—an extensive rather than intensive system—needs a support area of only 20 times the area of the mussel farm. Folke and Kautsky argued that intensive mariculture is not a substitute for capture fisheries, and the production of 33 million t from mariculture may not be sustainably compatible with the current production of marine capture fisheries.

Despite these problems, aquaculture does have the potential to reduce some adverse consequences of capture fisheries. For example, as mentioned above, Alverson et al. (1994) estimated that 9.5 million t—more than one-third—of the world's discarded bycatch resulted from shrimp trawling, although some shrimp are trawled to provide broodstock for the shrimp farms and some shrimp farms have serious adverse effects on local coastal (e.g., mangrove) environments. Whatever other ecological and socioeconomic consequences the culture of shrimp might have, to the degree that it reduces demand for wild-caught shrimp it is likely to reduce that bycatch. However, adverse environmental effects can include genetic and food-web consequences of genetically modified or nonnative organisms that escape (e.g., NRC 1996b); degradation or destruction of wetland and mangrove habitats to provide space for mariculture facilities; contamination of surface- and groundwater by fish wastes, pesticides, antibiotics, and other drugs; generation of red tides and related phenomena; and overfishing of wild populations to provide broodstocks for mariculture farms, in addition to the effects described by the NRC (1992b).

Clearly, a simple generalization about mariculture is not possible. Mariculture is practiced in many different ways, some of which have fewer environmental effects than others. In some cases, in situ culture of suspension-feeding organisms such as oysters can help remediate the effects of fishing by helping to restore an ecological function to the ecosystem (Lenihan and Peterson 1998). Nonetheless, it seems safe to say that the potential for present practices to supplement marine capture fisheries should be carefully evaluated wherever it is being used or proposed.

ENVIRONMENTAL CHANGE AND VARIABILITY

Marine ecosystems respond quite differently to environmental changes over various time scales, and they are different enough from terrestrial ecosystems that much of our understanding of terrestrial ecosystems does not apply to marine ones (Steele 1985, 1991a, 1991b, 1991c, 1996, 1998). Oceanic environmental changes occur diurnally (tides), over periods of several days (storms), over periods of months to years (upwelling, eddies, warm-core rings [Hofmann and Powell 1998]), over several years (El Niño-Southern Oscillation, North Atlantic Oscillation, variations in sea-ice cover in the Barents and Bering seas and Southern Ocean), and up to a century or more (the conveyor-belt, a large-scale ocean-circulation system involving all major oceans from arctic to antarctic latitudes [Broecker 1991]). In addition, human activity other than fishing has affected the marine environment and marine and anadromous fishes at a variety of time and space scales.

Natural[4] environmental changes have been implicated in several examples of changes in marine ecosystems, some of them described earlier in this chapter (e.g., the Bering Sea, the Barents Sea). Indeed, large fluctuations in population densities (or at least distributions) of marine fishes have been documented from periods long before fishing could have been a factor, most notably the very large fluctuations in the densities of scales of hake, anchovy, and sardine in sediment cores off California over the past 2,000 years (Soutar and Isaacs 1974) (Figure 3-1). Because they are so widespread and have been widely reviewed elsewhere (e.g., Wooster 1983, Wooster and Fluharty 1985, Laevastu 1993, Everett et al. 1996), only a few examples of environmentally related fluctuations in fish populations are provided here, both human caused and natural. We caution that it is often difficult to disentangle the effects of environmental changes from those of fishing on fish populations; often they are both important factors. The recent study by Polovina and Haight (in press) of spiny lobsters in a protected and an unprotected area in Hawaii is an excellent example of how experimental or observational controls are needed to prevent the effects of fishing and environmental changes from being confounded.

Salmon in the Pacific Northwest

Salmon populations and fisheries in the Pacific Northwest have declined severely over the last century (NRC 1996b). Salmon have disappeared from 40 percent of their historical breeding ranges in the continental United States and continue to decline despite a public and private investment of more than $1 billion in the last decade for protection and enhancement of salmon populations. The frequency of populations experiencing the greatest difficulty increases in a

---

[4]The term *natural* is used here to mean a phenomenon not directly caused by human activity (i.e., as a synonym for "nonanthropogenic").

southward trend from British Columbia to California. Species that spend extended periods as juveniles in freshwater are generally extinct, endangered, or threatened over a greater percentage of their ranges than species with abbreviated freshwater residence. In many cases, populations that have not declined in numbers are now composed mainly or entirely of hatchery fish, which tend to replace and otherwise threaten the gene pool of remaining wild salmon populations (NRC 1996b and references therein).

The salmon's life cycle begins in clear cold streams where salmon hatch, extends to the ocean where they grow, and returns again to natal streams where they spawn. Salmon habitats are degraded by agriculture, dams, forestry, grazing, industrialization, and urbanization, causing fish populations to decline. A recent NRC study (NRC 1996b) concluded that a strategy is needed to protect ecosystems and encourage the natural regeneration of lost salmon habitat.

Salmon productivity in the Pacific Northwest appears to be correlated with large-scale environmental changes in ocean circulation, temperature, and chemistry (NRC 1996b). These changes appear to be related to the ones that affect productivity in the Gulf of Alaska and Bering Sea and out of phase with them (NRC 1996a, 1996b, and references therein). Thus, when ocean conditions favor salmon productivity off California, Oregon, Washington, and southern British Columbia, they appear to be less favorable than average off northern British Columbia and Alaska, and vice versa. The interactions among fishing, human-caused environmental degradation in the streams where salmon spawn, and natural fluctuations in ocean climate are difficult to disentangle and complicate salmon management. Yet failure to consider natural fluctuations reduces the chances that management will be effective (NRC 1996b).

Upwelling Systems and ENSO Events

Upwelling systems have been reviewed by Everett et al. (1996). They are affected by larger-scale ocean-climate changes, such as El Niño-Southern Oscillation (ENSO) events, and changes in upwelling patterns change the distribution and abundance of many pelagic species (Bakun 1993). Upwelling is most common off west coasts of continents, as surface currents that flow towards the equator along the coasts are deflected away from them by the earth's rotation (Coriolis force). The water that flows away from the coasts as a result is replaced by colder, nutrient-rich water from below, and this upwelling supports abundant phytoplankton and hence fish production. ENSO events reduce the strength of the coastal currents, and hence the upwelling; as a result, coastal oceans off the west coasts of continents experience much warmer water than normal with fewer nutrients, less ecosystem productivity, and a variety of warm-water species that often are predators of or competitors with commercially important species. These ENSO-induced changes in upwelling patterns affect populations of small pelagic species such as sardines, pilchards, and anchovies off the coasts of Japan, Cali-

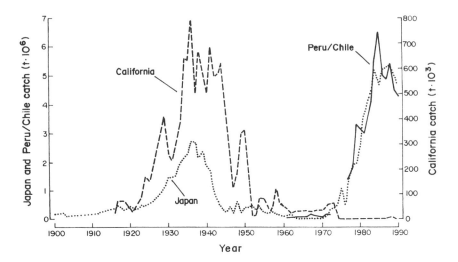

FIGURE 3-1   Record of catches of sardines in Chile and Japan, showing fluctuations, probably influenced by environmental fluctuations. Source: Redrawn from Lluch-Belda et al. (1989).

fornia, Chile, and Peru. Similar changes have been observed off west Africa (Bakun 1993).

ENSO events can have dramatic effects on fish populations. The 1972 event, combined with heavy fishing pressure, led to a major collapse of the Peruvian anchovy (Arntz 1986, Muck 1989, Sharp and McLain 1995). ENSO events, along with multiple impacts on their freshwater ecosystems, have also affected salmon fisheries off the Pacific coast of North America, depressing productivity and catches (Pearcy 1992, NRC 1996b). A strong ENSO event occurred in 1997-1998, and information on its effects on marine ecosystems and fish populations will increase our understanding.

Eutrophication

Eutrophication is a common and widespread human impact on the coastal ocean and in semienclosed seas, such as the Baltic and Mediterranean seas, often associated with other forms of pollution (Laevastu 1993). In estuaries and coastal areas, eutrophication and pollution can lead to algal blooms, including toxic dinoflagellates. Often, oxygen is depleted from benthic waters and fish kills result. However, eutrophication can increase productivity at various trophic levels, including fish. Examples of increased fish productivity attributed to eutrophication include the Baltic, where fish production increased substantially, partly because of eutrophication and partly because of reduced marine-mammal populations (Elmgren 1989, Nehring 1991). In the Mediterranean, increased pelagic fish production (e.g., anchovies) has been attributed to eutrophication as well (Caddy and Griffiths 1990).

MULTIPLE IMPACTS ON ECOSYSTEMS

This chapter has described the effects of fishing and environmental fluctuations as well as the effects of some other human activities on ecosystems. Because the impacts do not act alone, we discuss here three ecosystems—Chesapeake Bay, the Laurentian Great Lakes, and San Francisco Bay—that have been subjected to numerous significant human impacts and environmental changes. Although the Great Lakes make up a freshwater system and Chesapeake Bay and San Francisco Bay are semienclosed estuaries, some people have argued that their condition represents a preview of the fate of the larger, more open marine ecosystems discussed earlier in this chapter. Kerr and Ryder (1997), for example, described a history in which people continued to believe that the next larger ecosystem was so large that it could never suffer the damage of smaller, more contained ecosystems. The Laurentian Great Lakes are the largest freshwater ecosystem in the world, yet even they have been dramatically altered by human activities. Kerr and Ryder argued that coastal marine ecosystems, in particular those off Atlantic Canada, are next in line for similar changes; eventually the open oceans will be affected as well. Although San Francisco and Chesapeake bays are much smaller, they are connected closely to the open ocean, which affects them. Changes in the biota and physical environment of San Francisco Bay also have affected large areas of California's coast, as described later in this chapter.

This committee is unable to predict whether the events described in these three ecosystems are a preview of events to come in marine ecosystems in general, but it is clear that they are analogous at least for semienclosed and some other coastal systems. It is at least possible, unless significant changes are made in the way marine ecosystems are treated, that even open-ocean systems are at risk.

Chesapeake Bay

The Chesapeake Bay is the largest estuarine system in the United States, and its 166,000-km^2 watershed has supported important fisheries for centuries. Major changes in the bay ecosystem include eutrophication (from increased nutrient inputs), increased turbidity, decreased seagrass growth, and hypoxia in summer months. Fishery yields have remained surprisingly constant, or even increased, over the past 50 years, but the diversity and value of the harvest have declined as catches of many valuable species, especially oysters and anadromous fishes, have dwindled (Miller et al. 1996). Sturgeon, American shad, and two species of river herrings have declined precipitously in abundance. Seasonal migrants to the bay (e.g., bluefish and weakfish) also have declined, probably because of overfishing on the Atlantic Coast and within the bay. Catches of eastern oysters fell dramatically in the 1980s and 1990s, the victim of overfishing, habitat destruction, and parasitic diseases (Rothschild et al. 1994) and the interactions of those factors with altered water quality (Lenihan and Peterson 1998). In the summer of 1997, there were outbreaks in some Chesapeake Bay tributaries of a predatory form of the dinoflagellate *Pfiesteria piscicida*, which attacks fish. Because of fears that the dinoflagellate might also harm humans, a section of the Pocomoke River, a Chesapeake Bay tributary in southern Maryland and Virginia, was temporarily closed (Pesticide and Toxic Chemical News 1997). The reasons for the dinoflagellate's transformation to a toxic phase are not known, but most of the fish kills have occurred in estuaries with elevated concentrations of nitrogen and phosphorus (Burkholder 1998).

Not all species have declined in the bay. Striped bass populations collapsed in the 1970s (Richkus et al., 1992) but were placed under a five-year fishing moratorium and then recovered spectacularly in the 1990s (Leffler, 1993), a reminder that single-species management, if pursued vigorously, can be successful for some species. Spanish mackerel populations have increased coastwide; they have become much more common in the bay over the past five years than before. Other populations have proved remarkably resistant to damage from heavy exploitation by commercial and recreational fishers, most notably blue crabs (the dominant species in terms of harvest value) and menhaden (the dominant finfish species in terms of weight of catch), although the status of both species is being carefully monitored at present.

The Chesapeake Bay situation provides important examples of the need for an integrated ecosystem-based approach to management (Lenihan and Peterson 1998) as well as attempts to implement ecosystem management. The Chesapeake Bay Program (CBP) is the largest and best coordinated effort to restore a major coastal ecosystem in the United States. The goal of this federal-state program is to plan and implement total ecosystem management (EPA 1995a), with its highest priority being restoration of the bay's living resources (EPA 1995b). One objective of the CBP is to reduce controllable nitrogen and phosphorus inputs to the bay by 40 percent. If this objective is achieved, hydrodynamic and ecosystem models indicate that summer hypoxia would decline, benthic and deepwater habitats could be reclaimed, and seagrasses might be restored (Magnien 1987), returning the system to a more desirable state.

Management and rebuilding of living resources have been identified as goals by which the CBP's success will be judged. Fishery management plans for major exploited species have been developed jointly by the states in the watershed. To date, these plans are based primarily on single species, although some include aggregates of species, each treated individually. It is generally recognized that the plans are deficient with respect to habitat requirements. Although the plans may meet CBP goals for restoration of living resources, they fall short of meeting the program's call for ecosystem management. Multispecies management is now being explored. Whether multispecies management and ecosystem management are only distant goals or are imminently achievable are questions that the CBP will face as it attempts to reconcile the demands of multiple users with conditions required to restore a diverse, productive, and sustainable ecosystem.

The Laurentian Great Lakes

Great Lakes fisheries and the fish communities on which they are based have changed dramatically in the past 150 years. These large lakes contain one-fifth of the surface fresh water on the earth; their drainage basins are heavily developed and contain large portions of the human populations of the United States and Canada. Changes in fish communities have resulted from invasions and stocking of exotic species; overfishing; pollution; and loss or damage of habitat, especially in bays, tributaries, and shallower basins. The usefulness of fish as human food has been reduced by bioaccumulation of anthropogenic toxic substances; those chemicals have probably affected a variety of fish populations directly, especially salmonids (e.g., Mac and Gilbertson 1990, Mac et al. 1993). Yet important commercial (e.g., lake whitefish [*Coregonus clupeaformis*], Ebener 1997) and recreational fisheries persist or have recovered. Indeed, recreational fisheries have become very important in the Great Lakes, involving both introduced species such as Pacific salmon (*Oncorhynchus* spp.) and native species such as walleye (*Stizostedion vitreum*) (Knight 1997, Lichtkoppler 1997) and lake trout (*Salvelinus namaycush*) (Schreiner and Schram 1997). The management paradigm for the Great Lakes has been to take an ecosystem approach to rehabilitation of the ecological systems on which the fisheries depend (Francis et al. 1979). Even though the problems have been catastrophic at times, such as the elimination or near-elimination of fish species (Kerr and Ryder 1997), there is some positive sentiment among fishery managers, because of some successes in reversing or coping with a number of serious problems and because of the important recreational fisheries.

Sea lamprey (*Petromyzon marinus*), alewife (*Alosa pseudoharengus*), and rainbow smelt (*Osmerus mordax*) have invaded the Great Lakes in the past, severely affecting the native pelagic fish species. A combination of biocides to kill larval sea lamprey, fisheries for alewife and rainbow smelt, and artificial propagation of lake trout and Pacific salmon have been used to rehabilitate native populations and provide sport fisheries. Problems continue in terms of fish health (disease) and stocking at levels that overexploit the prey base, and the assemblage of fish species in the Great Lakes today is enormously different from the preindustrial assemblage. In particular, species of Pacific salmon, sea lamprey, and rainbow smelt are common, and some species have been exterminated (Becker 1983, Kerr and Ryder 1997).

Sources of input to the lakes of toxic and carcinogenic compounds have been much reduced and in many ways the Great Lakes ecosystems are recovering. Mercury is no longer at levels that are dangerous in Lake St. Clair. Dichlorodiphenyltrichloroethane (DDT) was banned, as were polychlorinated biphenyls (PCBs), and by 1995 both had declined in all of the lakes to concentrations below which regulations require action ("action levels"). Reduced input of phosphorus was achieved from metropolitan and industrial point sources and the use of low-phosphate detergents. Phosphorus levels have declined to action levels and water quality has improved in the past 30 years. Nitrates and nitrites seem to be increasing, however, and will be more difficult to control. Eutrophication has resulted in increased algal blooms, especially in Lakes Erie and Ontario.

Perhaps the greatest problem in the Great Lakes today is the rate of arrival of undesirable exotic species, primarily through transportation in ships' ballast water, but also through accidental introductions caused by other human activities and some deliberate introductions. Among recent invaders of concern are the zebra mussel (*Dreissena polymorpha*), the European river ruffe (*Acerina cernua*), and a predaceous water flea (*Bythrotrepes cederstroemi*). According to Mills et al. (1993), 139 species of nonindigenous plants and animals have been introduced into the Great Lakes; at least 13 of those species have had significant ecological effects. Future invasions might be reduced by revised methods of handling ballast water (NRC 1996d), but these and other exotic species probably will spread throughout the Laurentian Great Lakes and beyond.

The Laurentian Great Lakes are a primary example of the dependence of fisheries on the entire land-water ecosystem and the direct and indirect impacts that humans have had on fisheries. Problems will continue to develop and persist. The region has a better chance of dealing with these issues and problems as they develop because of the coordinating roles played by the Great Lakes Fishery Commission and the International Joint Commission on Water Quality. However, Kerr and Ryder (1997) argued that the Great Lakes experience is an ominous portent for Canada's Atlantic Ocean fisheries, which will follow the same pathways and will be harder to remediate than Great Lakes fisheries.

San Francisco Bay

The San Francisco Bay estuary is the largest such body on the Pacific Coast of the United States. Including its delta, it encompasses more than 4,100 km². Before 1950, the estuary contained 1,400 km² of freshwater marshes and 800 km² of salt marshes. The bay estuary drains 40 percent of California (Nichols et al. 1986). Runoff from the western slopes of the Sierra Nevada Mountains flows into the San Joaquin and Sacramento rivers and then into the San Francisco Bay and the Pacific Ocean. The estuary supports more than 120 species of fish and is critical habitat for migratory waterfowl (California Fish and Game Department 1998).

The estuary has undergone massive alterations through time in its physical and biological properties. More than 95 percent of the historic tidal marshes have been leveed and filled. These alterations have significantly reduced most native fish and wildlife populations. Most of the major rivers and streams that flow into the estuary have been dammed for flood control, power generation, and water supply. These structures, plus diversion canals, have reduced the inflow into the estuary by 40 percent, thereby altering flow and sedimentation patterns and water temperatures and blocking migration pathways for salmon and steelhead and altering their spawning habitats (Nichols et al. 1986).

Hydraulic mining in the foothills of the Sierra Nevada during the gold rush, especially between 1856 and 1887, resulted in massive inputs of sediments to San Francisco Bay. Starting in about 1950 with dam construction, some of these sediments were gradually lost, but there has still been a net sediment gain over the last century in the northern portions of the bay (San Pablo Bay) of nearly 400 million m^3. New input of sediments is now quite reduced because of upstream dams, and the waters are clearer than before (USGS 1998).

There continue to be large inputs of organic and inorganic chemicals into the bay. Toxic trace-metal accumulations accelerated during the 1950s. Some high accumulations of silver, cadmium, lead, and selenium are found at certain sites in the bay, which receives effluents from 46 wastewater-treatment plants and the discharges of 65 large industries. Approximately 40,000 tons of at least 65 contaminants accumulate in the bay each year. The sediments have been the repository of many organochlorine compounds, some of which bioaccumulate in the livers of striped bass and may be one cause of their declining populations (Pereira et al. 1994).

Sewage treatment plants discharge about 60 tons of nitrogen into the bay every day. Despite this input of fertilizer, the bay has not become more eutrophic because the very large populations of filter-feeding invertebrates keep phytoplankton from building up. Formerly, when ammonium nitrogen was not removed from treated sewage water, the south bay, with its poor circulation, became anoxic and fish died. Advanced sewage treatment has alleviated this problem and reduced the input of toxic metals (Cloern and Jassby 1995, Cloern 1996). Superimposed on this massive alteration of the San Francisco Bay estuary by human action are patterns of periodic major perturbations of the operation of the bay ecosystem in response to wide interannual variations in rainfall that affect salinity gradients. These perturbations are particularly evident during El Niño events (Peterson et al. 1995).

The Invaders

San Francisco Bay has been, and continues to be, considerably altered by invasive species. In recent years a new marine introduction has occurred about every 14 weeks. Many of these introductions have large impacts on the bay ecosystem. The Asian clam *Potamocorbula amurensis* first became established in 1986. In two years it became commonest clam throughout the northern part of San Francisco Bay (Carlton et al. 1990). It has reached very high population densities and altered the water chemistry of the bay and hence has affected many dependent organisms. In 1989, the green crab (*Carcinus maenas*) invaded the bay and is now spreading throughout California's coastal waters, where its voracious appetite threatens the shellfish and crab industries in the coastal regions. At present, there are 164 known introduced species of plants, invertebrates, and fish in San Francisco Bay, many of which have displaced native species (Cohen and Carlton 1998).

The invasive species are causing direct changes in the food webs of economically important species as well as altering the physical nature of the bay as noted above. For example, the major food item for fish, including juvenile striped bass (itself an introduced species), is the zooplanktonic mysid *Neomysis mercedis*. *Neomysis* is being displaced by introduced *Acanthomysis*.

Changing Nature of San Francisco Bay Fisheries

As might be expected from the ecosystem disruptions noted above, fisheries in the bay have had a tumultuous history (Box 5-4). Commercial fisheries for salmon, sturgeon, sardines, flatfishes, crabs, and shrimp were established soon after the start of the gold rush to support the rapidly growing human populations. These fisheries, and others that developed later, especially for striped bass, annually provided millions of pounds of protein until changes in the estuary severely reduced them. Loss of the fisheries can be attributed to a variety of causes, including overfishing, changes in water quality, and reductions in freshwater input through the bay delta.

Shellfish

The native oyster *Ostrea lurida* was intensively harvested beginning in the 1850s and gradually declined in abundance. Larger oyster species were imported and cultivated, first from the Pacific Northwest, and subsequently from the east coast (*Crassostrea virginica*). In 1899 more than 1,100 t of oyster meat was produced in the bay. The increasing pollution in the bay resulted in a decline of production by 50 percent by 1908 and by 1921 no more attempts at cultivating oysters were made. The bay was, however, used for holding imported stock until 1939, when the industry closed. Oyster culture has been moved to cleaner bays elsewhere in California (Leet et al. 1992).

Crustaceans

Bay shrimp, an aggregate of four species with *Crangon franciscorum* being the primary one, have been harvested commercially from San Francisco Bay since the early 1860s. By the 1890s, annual landings exceeded 2,300 t. Bycatch problems related to shrimping have led to fishing restrictions. Today the catch is down considerably from the peak years, and the shrimp is now sold mainly for bait (Leet et al. 1992).

Dungeness crab (*Cancer magister*) used to be commercially important in San Francisco Bay. Bay-area landings reached more than 3,500 t in 1956-1957 but have been less than 450 t during most years since. The fishing fleet in the bay area has been halved since 1957. The reasons for the collapse are not precisely known but probably include bay pollution, predation by nemertean worms, and changes in ocean temperature (Leet et al. 1992).

Striped Bass

A total of 432 striped bass (*Morone saxatilis*) from the Navesink River in New Jersey were deliberately introduced by the U.S. Bureau of Commercial Fisheries to San Francisco Bay in 1879 and 1882, along with American eels (*Anguilla americanus*), American lobsters (*Homarus americanus*), and other species (Lampman 1946). Although the lobsters and eels did not become established, the striped bass did, soon supporting major commercial and recreational fisheries. By 1899 more than 540 t of bass were commercially harvested (Skinner 1962). Striped bass have become established from approximately Monterey Bay, California, to Coos Bay, Oregon, with occasional fish being found as far south as northern Baja California, Mexico, and as far north as southern British Columbia, Canada (Hart 1973, Moyle 1976). The growing competition between recreational and commercial fisheries resulted in the banning of commercial fishing in 1935. Despite a gradual decline in striped bass populations, the sports fishery was still valued at $45 million in 1985. The populations in San Francisco Bay have continued to decline (California Fish and Game Department 1997), probably because of many factors, including water diversion, a changing food web caused by invasive species, and pollutants, although knowledgeable anglers continue to catch fish.

White Sturgeon

The commercial fishery for the longed-lived white sturgeon (*Acipenser transmontanus*) was quickly exhausted. During 1887, more than 680 t was caught. The catch dropped to 136 t by 1895, and the fishery was closed in 1917. Freshwater flow into the bay is apparently important for recruitment in this species as well (Leet et al. 1992).

Pacific Herring

Never a favored fish commercially in this region, Pacific herring (*Clupea pallasi*) has seen large fluctuations in population sizes in recent years. Its bay population was halved during the 1983-1984 El Niño event. At present, as discussed elsewhere in this report, an interesting and important industry of herring roe-on-kelp has developed in San Francisco Bay (Box 5-4, Leet et al. 1992).

Endangered Fish Species

Several fishes are endangered in the San Francisco Bay estuary. Populations of the delta smelt (*Hypomesus transpacificus*), found only in the estuary, plummeted in the 1980s. It was listed as threatened in 1991. The southernmost spawning populations of chinook salmon (*Oncorhynchus tshawytscha*) occur in the estuary. The winter runs of this species were listed as endangered in 1990. Management of these threatened species, within a complex ecosystem, were in part the motivation for the joint California-federal program noted below (Leet et al. 1992).

A New Start

As a result of management conflicts in protecting endangered species, meeting water-quality standards, and providing irrigation water through the Central Valley Project Improvement Act, a landmark framework agreement was formalized in 1994 to establish a state-federal cooperative agreement (CALFED) to provide solutions to the San Francisco Bay delta-estuary problems. In the long term this agreement would address the health of fish and wildlife as well as water-supply reliability and quality. An ecosystem-restoration plan is part of the program, which would work on "unscreened water diversions, waste discharges and water pollution prevention, fishery impacts due to harvest and poaching, land derived salts, exotic species, fish barriers, channel alterations, loss of riparian wetlands and other causes of estuarine habitat degradation"—in essence, an ecosystem approach to balancing various demands on the estuary. In addition to the federal and state agencies, a 30-citizen advisory group representing agriculture, environmental organizations, urban considerations, business, and fishing will be involved in fulfilling the goals of the framework agreement. This complex decision process is well justified to address this exceedingly complex management issue.

CONCLUSIONS

Fishing has had substantial ecosystem effects in most estuarine, coastal, semienclosed, and continental-shelf marine waters. It is likely that these effects are larger than the data suggest, for two reasons. First, in many cases the largest effects occurred before the ecosystems were carefully studied. Second, information is often lacking, even for well-studied systems. In addition, it is likely that in some systems the ecosystem effects of fishing have not fully developed or worked their way through the ecosystem. Although there is little information on ecosystem effects of fishing in the open ocean, those ecosystems that have been affected by fishing—estuarine, coastal, semienclosed, and continental-shelf ecosystems— provide most of the world's fishery products and many other services.

In addition to fishing, anthropogenic impacts and environmental changes continue to occur and are important. Anthropogenic impacts, which include contamination with toxic and other chemicals, habitat alteration and destruction, and introduction of exotic species, can be identified and in some cases managed. Most natural environmental changes cannot be managed in the traditional sense. Indeed, many are not even predictable in a precise way. However, they are becoming better understood in a general way and must be taken into account in any sensible management program.

4

Diagnosing the Problems

The preceding chapters have established that many marine fisheries are over-exploited and that fishing has adversely affected many marine ecosystems. At the most basic level, the problem is that too many people are catching too many fish, sometimes in the wrong place at the wrong time, and sometimes using equipment and techniques that damage natural ecosystems. The world's fishing capacity greatly exceeds what is needed to catch the sustainable yield. Finally, with a growing human population and increasing industrialization, there is growing pressure and ability to catch fish. This diagnosis highlights the need for an analysis of the major factors that have contributed to the current state of affairs. In broad outline they can be divided into three major categories: scientific matters, management matters, and socioeconomic incentives.

SCIENTIFIC MATTERS

Scientific matters can be further divided into two major categories: lack of adequate scientific information and failure to use existing scientific information appropriately. We begin with the information itself and then discuss its use.

Assessments of Stocks and of Fishing Mortality

A fundamental premise of fishery management is that the productive potential of a stock and a fishery on it is a function of the abundance and biomass of the animals present in the stock and their life-history characteristics. These characteristics include the age distribution, natural mortality rate, age at maturity, and fecundity as a function of age. Most fishery-management programs depend to a degree on assessments of the stock and its productive potential (NRC 1998a).

The primary purpose of a stock assessment is to determine the abundance of individuals in the stock. In addition, fishery scientists estimate the mortality rate

64

and partition it into components from fishing and natural causes, allowing the exploitation rate to be calculated. Based on the stock's growth potential and mortality rate, its productive potential is estimated. Usually, estimates of the fishing mortality rate and stock size that can optimize (or maximize) the catch on a sustainable basis are provided. Thus, a determination of the status of the stock relative to present fishing intensity and the stock's ability to sustain additional fishing are two outputs of an assessment.

Assessing a fish stock is an inherently difficult problem. Except for anadromous fish (like salmon) migrating up rivers, it is impossible to actually count individuals, requiring that sampling methods be applied. They include fishery-dependent samples and data obtained from the fishery itself and fishery-independent samples from various survey techniques (e.g., trawl surveys, acoustics sampling, mark-recapture experiments). The sampling data usually provide the information that goes into the stock assessment. Because it is impossible to count all individuals in the stock, there is no perfect method to confirm the accuracy of the assessment techniques. Indeed, there are many biases: fishing and sampling gears selectively catch fish of certain sizes and behaviors. And the overall vulnerability or availability of fish to sampling gears can change with size and age (Ricker 1975; Gulland 1983). Fisheries that have been established for many years are difficult enough to assess, but recently established fisheries are even harder because long-term data are lacking.

A common problem in assessment is the nonuniformity of *catchability*, the probability that an individual will be caught by a unit of fishing gear. In schooling species such as herring or krill, or very large species such as whales where a single individual is worth pursuing, the relationship between catch and fishing effort is usually nonlinear (Ulltang 1980, Csirke 1988, Miller and Hampton 1989, Gulland 1983). In such fisheries, experienced fishers often can maintain high catch rates when the stock is declining in abundance. Under that circumstance, fishers—and sometimes fishery managers—can underestimate the mortality rate of the stock, making it easier to severely overfish it before regulations are instituted. This explains, for example, why a highly mobile and technologically sophisticated offshore fleet was able to maintain high catches of northern cod off Newfoundland while the less-mobile inshore fleet could not (Neis 1992, Steele et al. 1992, Finlayson 1994, Chapter 2 of this report). Walters and Maguire (1996) identified this problem as one that led to inaccurate stock assessments of northern cod, and indeed it is a reason that fishery-independent surveys are preferred in stock assessments.

Estimating fishing and natural mortality rates is also difficult, in large part because there is no perfect way to estimate abundance. Discards, mortality of escaped animals, and unreported catches are difficult and expensive to estimate precisely. However, they are part of fishing mortality, and the inability to estimate them accurately can lead to underestimates of fishing mortality, as discussed in more detail in Chapter 3 and in Chapter 2 for northern cod. Natural mortality, the sum of all mortalities not a consequence of fishing, also is difficult to determine and contributes significantly to inaccuracies in assessments. None of the above invalidates stock assessment as a fishery-management tool. However, in any industry where profit is a relatively small percentage of investment, as it usually is in fishing, it is difficult to forgo those small profits by reducing or stopping catches when there is a chance, but not a certainty, that the assessment indicates overfishing. The uncertainty cannot be reduced to zero, or even close to zero. Thus, the basic scientific uncertainty that inevitably accompanies estimates of stock size (abundance), productivity, and fishing mortalities is a reality that must be taken into account by any sustainable management program.

Environmental Variability

Environmental variability cannot be precisely predicted, and that leads to scientific uncertainty. Many fish populations fluctuate substantially from year to year and from decade to decade. In some cases these fluctuations are related to environmental fluctuations (Chapter 3); in other cases the causes are poorly known or unknown. Whether or not their causes are known, the fluctuations are often not predictable more than a year in advance, if that long. This unpredictable variability results in a rather fundamental scientific uncertainty: the future size of a stock is often unknowable, although probability distributions of future stock sizes are often estimated. Fluctuations in the population sizes of many commercially important marine species can occur at much shorter time scales than typical responses of the fishing industry, whose responses often depend on processes (such as shipbuilding, repayment of loans, and vocational training) with much longer time scales. The natural variability of many marine stock sizes interacts with uncertainties in current stock assessments to make precise planning impossible.

Landing Statistics

In some fisheries, landing statistics (the number, kind, and size of fish landed) are fairly accurate; in others they are less so. Even when landing statistics are accurate, however, they can bear an uncertain relationship to the number of fish killed by the fishery (Chapter 3). These uncertainties contribute to uncertainties in assessments of stock size, fluctuations, productivity, and fishing mortality. A fishery that tries to extract the last "surplus" animal (i.e., one that tries to maximize yield not over the long term but each season) is flirting with danger: many of the uncertainties described above work to deplete the fishery rather than to increase it. Despite the excellent work of many biologists in fishery agencies and universities, there will always be scientific uncertainty concerning how heavily a fishery can be exploited.

Uncertainties Concerning Socioeconomic Information

The uncertainties associated with landing statistics were mentioned earlier. Some uncertainty derives from uncertainties and lack of good information about human behavior. Indeed, the history of fishery management is full of examples of surprising human reactions to regulations—reactions that would not be surprising if managers had better and more detailed information. Limitations on boat lengths, for example, can lead to grotesquely wide boats, as in Bristol Bay (Alaska), where salmon boats are restricted to 32 ft in length, but have no restrictions on their beam. In Alaska, fishers responded to fleetwide halibut quotas not by reducing their fishing power but by fishing harder, so they could catch as many fish as possible before the overall quota was reached. That made the open seasons shorter and shorter until a "derby" fishery resulted, with hundreds of boats racing for fish in openings that ended up lasting less than a day (NRC 1994c, Buck 1995, Pennoyer 1997). A system of individual quotas was implemented in 1995, largely because of this distortion of fishing effort (Pennoyer 1997).

Despite considerable study, there remain major uncertainties in how fishers, their communities, and markets will respond to such management options as individual transferable quotas (see e.g., McCay 1995a), community development quotas, comanagement programs, international treaties, and so on. More needs to be learned about how fishery scientists and managers respond to uncertain information and to information that does not fit their scientific paradigm, such as traditional knowledge (Neis 1992, Finlayson 1994). The uncertainties regarding less-industrialized countries are even greater than for countries in North America, Europe, and Australasia. Mariculture has also introduced uncertainties by affecting the prices of wild-caught products, usually depressing them. Technological innovations also can have profound effects on fishing and the behavior of fishers, processors, and marketers. For example, the connection of Seattle to the east coast of the United States by the transcontinental railroad in the late nineteenth century provided a market for Pacific halibut, which resulted in an enormous increase in fishing pressure (Thompson and Freeman 1930). Other technological innovations, such as onboard freezers, also affected halibut (Bell 1978) and other fisheries (e.g., NRC 1992a) by allowing ships to stay at sea with their catches for weeks instead of only a few days.

How Scientific Information Is Used

Because scientific information concerning fisheries is to some degree uncertain, there is always a temptation to assume the best and treat the fishery as though the uncertainty will work to benefit rather than hurt the fishers (Ludwig et al. 1993). Thompson's (1919) insight that scientific information will have to be overwhelming to change sport and commercial practices remains as true in 1998 as it was in 1919. The fishery literature is replete with examples of misuse or even lack of use of scientific information. For example, shrimp trawlers off the U.S. southeast and Gulf coasts long rejected the conclusions of the National Marine Fisheries Service that they were largely responsible for killing endangered sea turtles (NRC 1990). Because the information contained some degree of uncertainty, the shrimp trawlers were able to resist attempts to use turtle-excluder devices in their trawls until the National Research Council's reanalysis of data clarified their contributions to turtle mortalities. The uncertainty does not always work against the fishers. Recently, the International Pacific Halibut Commission increased the allowable catch of Pacific halibut because it learned that its stock assessments had been too pessimistic: apparently, there are really more halibut than the stock assessments had indicated (Ana Parma, IPHC, personal communication, 1997). Knowledge that uncertainties can occasionally benefit the fishers by leading to more fish than were expected makes it that much more difficult to routinely forgo catches in the face of uncertainty. Many cases of overexploitation of fishery resources result from this cause. NMFS (1996a), for example, described many examples of U.S. fishery resources that have been "excessively fished" for many years. In other words, those resources have been exploited at higher rates than the scientific information—widely disseminated and never seriously questioned—supported, and no amount of scientific information would have changed this outcome.

Sometimes information is not used by policy makers and other stakeholders because it is not communicated to them in a way that is relevant and understandable. An important factor in communicating scientific information to managers and the public is to acknowledge and account for differences in the cultures of scientists, managers, policy makers, and the public. Disciplinary barriers, differences in operational constraints, and institutional differences are obstacles to good communication. There is a great deal to be learned from the experience and science of risk communication (e.g., NRC 1994f, 1996c).

MANAGEMENT MATTERS

Fishery management as a whole process (i.e., not necessarily individual fishery managers) is frequently blamed for failing to deal with the uncertainties in scientific information described above and for failing to take a conservative, or risk-averse, approach. But the fishery-management process includes many actors outside the management institutions themselves. In addition, fishery-management institutions often operate at time and space scales that do not match those of an individual fishery (e.g., NRC 1996a, 1996b). For example, the population fluctuations of some species occur so quickly that they cannot be determined until well into the fishing season, but most management and industry responses take longer than a season, especially those involving significant capital investments, such as boats and gear. In other cases, environmental variations affect the stock size of target species (Chapter 3) in often unpredictable ways. In many cases the target species ranges over very large areas, often crossing several management jurisdictions and sometimes several nations (e.g., NRC 1994b, 1994c, 1996a, 1996b), yet the jurisdiction of fishery-management institutions often reflects political boundaries. In other cases a watershed (NRC 1996b) or a large reef or bank might be the appropriate management unit. Progress has been made in dealing with such difficulties (Chapter 5), but responding at the appropriate time and space scales remains a challenge for management.

Another problem is that many managers are trying to balance diverse, even conflicting, but unarticulated goals (Rothschild 1983, Pikitch 1988, Policansky 1993b, Hutchings et al. 1997). Another aspect of this problem is that a variety of political agendas and potential conflicts of interest complicate fishery management (NRC 1994c, 1996b). The challenge of making fishery management an inclusive yet balanced and fair process is a daunting one, as discussed in Chapter 5. There is a tendency to think of fishers as a monolithic group, all intent on taking as many fish as possible, but this oversimplified view is not realistic. Even in small fisheries there are usually a variety of sectors with different goals and interests. For example, in the northern cod fishery, the inshore and offshore fisheries are very differently constituted and have very different interests (Finlayson 1994). Processors can have very different interests from fishers. Recreational anglers—especially in the United States—are another important interest group. Even within recreational fisheries, there can be an enormous diversity of goals and interests, as described by Merritt and Criddle (1993) for Kenai River chinook salmon (*Oncorhynchus tshawytscha*).

Multiple fisheries on a single species or population and single fisheries on multiple species or populations (mixed-population fisheries) immensely complicate managers' difficulties. For example, Pikitch (1988) suggested that maintaining the structure of a community unchanged might not be compatible with any catch rate. As another example, Pacific halibut are managed by the International Pacific Halibut Commission, but approximately one-quarter of halibut landings in 1990 were taken as bycatch in other groundfish fisheries (Thompson 1993). That situation complicates the management both of halibut and of the other groundfish, whose capture can be severely limited by restrictions on the halibut bycatch.

The problem of mixed-stock or mixed-species fisheries is that some species and stocks (populations) are more productive or less susceptible to fishing (catchable) than others, so if fishing pressure is low enough to protect the least productive or most catchable population, others are "underharvested" and there is great pressure to allow an increase in catch rate. When the catch rate increases, the less-productive populations (NRC 1996b) or species (Pikitch 1988, Roberts 1997) or more catchable populations (Clark 1990) or species (Brander 1981) are depleted. This problem can be quite serious; it appears to have contributed or led to the loss of species from some environments (e.g., some salmon populations or species in streams in the Pacific Northwest [NRC 1996b]) and might even cause the local extinction of a species, for example the common skate (*Raja batis*) in the Irish Sea (Brander 1981); the barndoor skate (*Raja laevis*) may be near extinction throughout its range in the Northwest Atlantic (Casey and Myers 1998). But the precise nature and timing of the interactions related to multispecies fisheries are very hard to predict (Pikitch 1988).

Finally, many fishery-management agencies have mandates and goals that are potentially in conflict. They are often asked to promote fishing and the fishing industry and to protect the ecosystem and the individual species in it. Sometimes, a goal—often unspoken but occasionally explicit (e.g., Task Force on Atlantic Fisheries 1983)—is the preservation of a fishing community's way of life. How fishery management affects fishing communities is a major issue all over the world. In the United States, the Sustainable Fisheries Act of 1996 requires fishery impact statements that assess the likely effects of management measures on fishing communities. The difficulty of dealing with goals that are not made explicit—much less agreed on by most of the parties involved—is common to many resource agencies, not only fishery agencies. While it can be dealt with, it often does not receive the attention it deserves.

Enforcement

Enforcement is often a difficult problem for fishery managers and is related to many of the scientific uncertainties described above (involving biological and social sciences). The incentives to bend the regulations or to cheat are many, and there are so many participants in most fisheries that it is impossible to prevent or catch all violations. Sometimes the regulations themselves are confusing or not well disseminated, resulting in unintentional violations. Recreational fisheries are particularly difficult to monitor, and to some degree they depend for compliance with regulations on an honor system. The problem is well known, and because it involves illegal activities, solutions are made more difficult. We provide one example, that of whaling.

International whaling is controlled by two organizations, the International Whaling Commission (IWC), which allocates catch limits, and the Convention on the International Trade of Endangered Species, which regulates the import and export of endangered species. Both organizations tightly restrict the hunting and trade of all baleen whales plus sperm whales. The IWC adopted a moratorium on commercial whaling in 1986 (Marine Mammal Commission 1998), although some species like the humpback whale and the blue whale have been fully protected since the mid-1960s. About 500 to 600 whales are taken each year for research or aboriginal use, the bulk of which are minke whales taken under scientific permit to Japan. The purpose of the IWC is to sustainably manage the whaling industry and as such it represents an important example of international fisheries control for the benefit of sustainable exploitation of global natural resources.

Whale meat is important commercially in Japan, where prices range up to $150 per kilogram. Until recently, it has not been possible to compare the makeup of the retail market with expectations based on IWC rules because most whale products are unidentifiable once they reach local markets (e.g., sliced bacon, dried marinated whale jerky, canned meat). However, DNA sequences can now be used to identify the species of whale on international markets (Baker and Palumbi 1994, 1996), making it possible to evaluate compliance of the retail market with international expectations. These DNA results show that enforcement of international regulations is too lax to adequately protect the world's whale populations. About half of the whale products on the Japanese market are from the expected minke whale populations. The other half is made up of unprotected species of small toothed whales (dolphins, beaked whales, porpoises) and prohibited baleen whales (Baker et al. 1996). To date, most of the world's baleen whales have been found in the retail market, including humpback whales, blue whales, fin whales, Bryde's whales, and northern minke whales (legal only since 1994). Blue whales are particularly threatened, with an estimated worldwide population of 4,000 animals, yet they continue to be part of the retail market.

These animals are entering the market under the cover of legal products, but they are probably taken by illegal whaling and shipping. Recently, information from the former Soviet fishing fleet has shown that tens of thousands of whales were taken illegally in the 1960s and 1970s without notice by the international community (Yablokov 1994). Whale-meat smuggling may be a lucrative business, and reports of confiscation of illegal whale shipments surface regularly (Baker and Palumbi 1996).

International agreements have led to recovery of many of the world's whale populations. However, enforcement of existing regulations is hampered by the scale of the oceans and the complexity of international shipping. Illegal activities dilute efforts to manage whale populations scientifically and threaten the balance of national priorities on which international agreements are based.

SOCIOECONOMIC INCENTIVES

The problems of the socioeconomic incentives in fisheries have been widely discussed. One problem is uncertainty, which can lead to risk-prone behavior. Another related (and better-known) problem is that of the "commons," where a lack of clear property rights leads to a difference between individual and short-term interests on the one hand and societal and long-term interests on the other. The classic statement of this "open-access" problem came from an analysis of fishing (Gordon 1954): it is in the interest of an individual fisher to increase catch, but it is not necessarily in the interest of the whole fishing community. Other problems, such as overcapitalization, derive from this one. A further problem is that some incentives, such as those derived from personal and societal goals, culture, and lifestyle, are not easily identified or incorporated into the bioeconomic models often used to develop fishery-management plans.

The conflict between individual short-term goals and long-term broader societal goals is strikingly illustrated by the plight of poor hungry people who depend on fishing for food. People in such straits cannot afford to worry about food tomorrow or food for others in the community if they do not get food today. However, even though they cannot avoid the strategy of satisfying today's needs at the expense of future needs, their plight will only get worse without some kind of intervention, because the fisheries will have less and less capacity to produce food. Some intervention is required to develop sustainable fisheries in such cases.

Economic Considerations

The problem of overfishing described in preceding chapters lies not so much with an inherent "greed" on the part of fishers but because, in most fisheries, fishers have faced an economic incentive system and a regulatory regime that lead almost inevitably to overexploitation and economic waste. Fishery resources are a form of natural capital. They can be seen as assets that can yield a stream of economic returns to society (Clark 1990, Clark and Munro 1994). They differ from human-made capital in that we receive an initial endowment of the capital assets from nature. It is possible to invest and disinvest in natural capital just as in human-made capital. Refraining from fishing and enhancement activities are investments in the resource. Fishing in excess of sustainable yield, and thus depleting the resource, is disinvestment.

Economic motivations for investments and disinvestments in fishery resources are similar to those for what might be termed conventional capital (e.g., plant and equipment). Investment in the resource requires a current sacrifice—deferred profits or costs of enhancements in the hope of a future return—while disinvestment provides immediate economic returns at the cost of lower future returns.

A key factor in investment decisions is the rate at which the investor discounts future economic returns as compared with current returns. The lower the discount rate, other things being equal, the greater will be the incentive to invest; higher discount rates lead to a lower incentive to invest. (The relative value of money today as compared with its value at some future time is known as the *time preference* of money and it is measured by the *discount rate*. Discount rates can reflect objective estimates of known relationships—e.g., depreciation of equipment, inflation rates, or the knowledge that if one doesn't fish soon, in a competitive-allocation fishery there might be no fish later—or they can reflect subjective time preferences.)

The payoff from the investment is enjoyed in the future, so resource investment, like other investment, involves uncertainty about the future. As a result, most individuals and societies give less weight to (risky) future than to current returns. In other words, risk is one cause of the future's being discounted, which reduces the incentive for investing. If the risk to future returns is high enough, the incentive to invest can disappear and disinvestment will occur. These changes caused by the presence of a positive discount rate were not considered explicitly by Gordon (1954) in formulating his concept of the economically optimal fishing effort that results in maximum economic yield. Gordon implicitly assumed that the appropriate social discount rate is zero, i.e., that future economic returns from the resource should be given the same weight as current returns (Clark and Munro 1982).

The above has made capture fisheries difficult to manage in economic terms, mainly because of poorly defined property rights (Gordon 1954). Even when coastal states claim property rights to fishery resources in their waters, it is often difficult or expensive to vest property rights in the resources to fishers on an individual or collective basis (Gordon 1954, Munro and Scott 1985). Fish are mobile and not easily observable before they are caught. This contrasts with agriculture, in which property rights to the basic natural resource—land—are well defined.

The poor definition of property rights has two major consequences. First, individuals have a strong incentive to discount future returns heavily and not to invest (Clark and Munro 1982). If they attempt to invest by not fishing, they might do no more than increase their competitors' catches. Instead, the incentives lead them to increase their own share of the available resource by fishing more, rather than less. The second major consequence, which follows from the first, is fleet (and perhaps processing) overcapacity. These conditions—increased fishing effort and increased capacity—make overexploitation more likely.

Practical Considerations for Management of Fisheries

In practice, fishery management has two critical elements. The first—related to conservation—is intended to limit fishing mortality and is implemented through input controls and output controls.[1] The second can be explicit, or implicit or by default, and that is related to the processes by which access to the resources are allocated. The default situation usually involves no specific action. If managers attempt to prevent overexploitation by limiting fishing effort without changing a competitive allocation scheme, the limited catch will become a common pool. Each fisher will be competing with others for a share of the total resource, which is now limited. This competition often leads to increased capital investment in fishing effort (gear, boats, human resources, and so on), a phenomenon called *capital stuffing*. Soon there is more fishing capacity than needed to catch the limited resource, which leads to a race for the fish. In addition, as regulations are successful in increasing the size of the resource (increasing fish populations), profits from fishing will increase and make additional investment

---

[1]Input controls, by analogy with other production systems, are controls of expenditures on catching fish, such as boats, gear, fuel, and such items. Output controls control the product of fisheries, i.e., the catch.

worthwhile, which also results in overcapitalization. The investment of individuals in more fishing capacity is entirely rational, even though the total catch will not increase.

The primary alternative to a default or competitive allocation process is share-based or rights-based allocation. This approach also can provide incentives for conservation because participants (rights- or share-holders) have a stake in the future of the resources and because some rights can provide incentives for efficiency (output reduction). The promotion of efficiency occurs because in the absence of competition for shares of the resource, economic success is represented by fishing efficiently. These are the main reasons why this committee has encouraged the development and use of share-based allocation systems to replace competitive allocation schemes.

The following paragraphs briefly describe some of the theory and experience of fishery management from an economic viewpoint. For more detailed discussions, the reader is encouraged to read Clark (1990) and OECD (1997) and references therein. The OECD publication in particular contains many recent references and ample examples of fishery management in practice.

In theory, the imposition of conservation measures strong enough to be effective, either through input controls such as gear limitations or seasonal and areal closures, or through output controls such as catch limits (usually total allowable catches or TACs), allows allocation methods to be considered independently of conservation measures. In practice, conservation and allocation methods can become dependent if competitive allocation drives up fishing costs to the point where rent is dissipated or marginal. The management of Pacific halibut before the implementation of ITQs is perhaps the best example of this: the resource was protected for decades, but the race for fish became excessive and dangerous with many adverse social and economic consequences, as described earlier in this chapter.

If the system were well balanced, or theoretically perfect, the above might not be a problem; when rents approached zero, investment would decline to reflect that. But in practice, three important factors cause problems. The first is imperfect information about the current and future size of the fishery resource, which can lead to costs (inputs) that exceed what the resource can sustain. The second factor is natural variability, which can have a similar effect. The third factor is the cost of complying with regulations (for example paying for and using specified gear modifications). The costs of compliance as well as of licenses or other fees can increase the costs of fishing, which can affect profitability, especially if other suppliers of the market do not bear the same costs.

The first two factors are particularly problematic when a fishery develops on a resource that has a high standing stock. In that case, the investment is often made in response to the standing stock, analogous to financial capital, rather than in response to the productivity of the resource, analogous to return on investment. In such cases one observes a so-called *ratchet effect* (Ludwig et al. 1993) and the investment overshoots the appropriate level for the productivity of the resource and excess capacity is a result.

Excess capacity is a form of economic waste and makes a fishery vulnerable to resource shocks (e.g., reduced availability of fish or regulations to reduce catch) or to economic shocks (e.g., falling prices or increasing costs, such as happened in the 1970s with the oil embargo). Excess capacity is difficult to estimate quantitatively, but a few quantitative and many qualitative estimates—ranging as high as 75 percent for some fisheries—were described by Mace (1997). For Alaska groundfish fisheries, Pennoyer (1997) estimated that overcapacity was 300 to 400 percent before the introduction of individual quotas. Often, the vessel and processing capital cannot be used for purposes other than the specific fishery. Vessel and factory owners usually have debts, such as mortgages, to be serviced. As the excess capacity dissipates economic returns from the fishery, the incentive for fishers and processors to press for liberal catch quotas increases as does the political influence of the users: if their catches are significantly reduced, they often face bankruptcy. The industry often uses scientific uncertainty or natural variability to argue against reduction of effort (OECD 1997). The pressure for liberal catch quotas can be very strong—often involving important political figures—and risk-prone management often results (Sissenwine and Rosenberg 1993, Rosenberg et al. 1993). Even if managers resist pressures to make risk-prone decisions, the existence of a large, chronically undersatisfied fleet exacerbates monitoring, control, and surveillance (Dupont 1996). If the fishing capacity of the fleet is held to levels at or below that required to produce the maximum sustained yield or the maximum economic yield, then the possibilities of overfishing are substantially diminished (OECD 1997). The more the fishing capacity exceeds that necessary to produce the desired sustainable catch, the greater the potential for overfishing. Indeed, overcapitalization is often cited as the most important factor in overexploitation of the world's fisheries (e.g., NMFS 1996b, FAO 1996b, Christy 1997, Mace 1997, WWF 1997). Nonetheless, there are examples of fisheries that have extreme excess capacity in which strong management has prevented overfishing even in the absence of share-based allocations (e.g., the Pacific halibut fishery before implementation of ITQs).

These practical matters cause pressures that often lead to or exacerbate overexploitation. If there is political will, they can be dealt with, but often that will has been lacking. No method of reducing fishing mortality to achieve conservation and thus sustainable fishing will be economically or socially painless; financial investments and jobs will be lost. However, if sustainable fishing is to be achieved, reducing effort in the short term is necessary. The options lie in deciding how and when to reduce effort so as to reduce economic and social disruption. The options, however, can be exercised only if decisions are made before the resources are depleted.

Subsidies

The problem of overcapitalization, which has been described in this report as a serious contributor to overfishing, is probably seriously aggravated by government subsidies. Estimates of the total subsidies in marine fisheries are extremely hard to obtain and vary widely, from as much as $46 billion annually to $11 billion annually (FAO 1993b, Garcia and Newton 1997, Milazzo 1997, Porter 1997). All authors agree that the estimates are very imprecise. Even the lower estimate, however, is very large, especially considering that the total gross revenues of the world's marine fishing fleet are estimated at about $70 billion (FAO 1993b) or $80 billion (Milazzo 1997) per year.

A recent symposium in New Zealand on fisheries and international trade devoted substantial time to subsidies and concluded that subsidies in fisheries are large and pervasive and that the motivations and impacts of the subsidy programs are variable and poorly understood (Pacific Economic Council Task Force on Fisheries Development and Cooperation 1997). Until our knowledge of these matters improves, it is difficult to make specific recommendations for dealing with them, although it is clear that they complicate attempts to reduce fishing effort by reducing overcapitalization (Porter 1997, Milazzo 1997); it is also clear that much additional research is needed. Milazzo (1997) suggested that environmental subsidies (e.g., vessel and fishing permit buybacks, stock enhancement, research and developments in "clean" fishing gear, perhaps others) are preferable to "conventional, effort- and capacity-enhancing" subsidies and should be given higher priority. The design and implementation of environmental subsidies need improvement, and certainly research is needed to understand their potential better.

CONCLUSIONS

A great number of scientific, management, and socioeconomic uncertainties and difficulties contribute to the overexploitation of marine fisheries and their ecosystems. Fisheries science cannot provide precise estimates of fish abundance or of the impacts of fishing, and the information that science can provide is not always well used. Environmental variations introduce a great deal of variability in fish populations and uncertainties in managing them. Pervasive and powerful economic and social incentives lead to overexploitation of fisheries. Some improvement is possible in each of these three areas, but it seems certain that the kinds of incremental improvements that have characterized recent decades will not by themselves reverse the trend toward increased overexploitation, although maintaining progress in those areas is essential. Creating more appropriate incentive systems and developing management institutions that can accept and deal with variability and uncertainty are crucial to establishing sustainable fisheries. Without them, populations of individual species and the structure and functioning of marine ecosystems are likely to continue to decline.

5

Options for Achieving Sustainability

Previous chapters have described the status of marine fisheries and ecosystems and identified some of the factors that have led to the current situation. This chapter discusses options for improving the prospects for sustaining marine fisheries. We begin with management and socioeconomic incentives because the committee believes that changes in those areas can have the largest and most immediate positive effects. We conclude with scientific considerations, many of which involve research. It is, of course, impossible to neatly categorize the following discussions as focusing on management, scientific, or socioeconomic matters. Many of the approaches discussed below include elements of all of them.

MANAGEMENT

Previous chapters have suggested that management difficulties include a lack of scientific information; a lack of full appreciation and use of available scientific information; a risk-prone approach; a lack of appreciation for ecosystem and other nonfishery values; the need to balance many goals and values, some of which conflict and many of which are not clearly articulated; and space and time scales of management that do not coincide with the distribution of the target species, their ecosystems, or fishing communities. The committee concludes that to approach the goal of sustainability managers should adopt a conservative, risk-averse approach that recognizes ecosystem values. After describing the U.S. management context, the focus below is on approaches most likely to be helpful in achieving that goal.

U.S. Management Context

Marine-fishery management in the United States takes place in several interrelated ways. They include management by states for stocks found largely within the 3-nautical-mile territorial sea; by the federal government through regional councils for stocks found largely in the exclusive economic zone (EEZ) 3 to 200 nautical miles from shore; and through international bodies for certain shared or highly migratory species such as Pacific halibut, Pacific salmon, tunas, and whales. In addition, there are some highly localized systems such as town management of oysters and clams in parts of New England as well as regional systems such as the interstate marine fisheries commissions (e.g., the Atlantic States Marine Fisheries Commission). The EEZ management to some extent sets the framework for other management regimes. It was established under the authority of the Magnuson-Stevens Fisheries Conservation and Management Act (MSFCMA) of 1976, reauthorized and amended most recently in the Sustainable Fisheries Act of 1996. The primary purposes of the act are to:

1. Establish a geographic zone adjacent to the United States over which the U.S. government is responsible for fishery resource management, with limited exceptions.

2. Promote conservation and achieve optimum yields from the nation's fishery resources. Social and economic factors are to be given equal importance for modifying optimum yield.

3. Create a legal and economic environment that stimulates harvest of fisheries resources within the area of extended jurisdiction, and subsequent processing of such catches by U.S. fishermen and companies.

4. Establish an institutional structure and enforcement authority that allows the United States to carry out the objectives explicit and implicit within the Act.

5. Ensure that conservation and management under the act are based on the best scientific information available (P.L. 94-265).

The organizational structure set up by the MSFCMA is based on eight regional fishery management councils, with representation from relevant state and federal agencies as well as the public. Public members have backgrounds ranging from commercial or recreational fishing to research and fishery conservation. The councils prepare fishery-management plans (FMPs) or plan amendments. They are implemented by the National Marine Fisheries Service (NMFS), if approved by the Secretary of Commerce. The NMFS also provides most of the data and stock assessments used in management.

Congress has amended the MSFCMA to revise National Standard 1 (Box 5-1) to require greater conservation; no longer can any relevant economic or social factor be used to justify fishing at levels above the maximum sustainable yield. The amendments also call for a reduction in bycatch and overcapacity and for more attention to habitat protection, specifically requiring the designation of essential fish

BOX 5-1
National Standards from the Sustainable Fisheries
Act of 1996, P.L. 104-297 (Amendments to Magnuson-Stevens
Fishery Conservation and Management Act of 1976):

SEC. 106. NATIONAL STANDARDS FOR FISHERY CONSERVATION AND MAN-
AGEMENT

a. IN GENERAL - Any fishery management plan prepared, and any regulation promulgated to implement any such plan, pursuant to this title shall be consistent
with the following national standards for fishery conservation and management:

1. Conservation and management measures shall prevent overfishing while
achieving, on a continuing basis, the optimum yield from each fishery for the
United States fishing industry.

2. Conservation and management measures shall be based upon the best scientific information available.

3. To the extent practicable, an individual stock of fish shall be managed as a unit
throughout its range, and interrelated stocks of fish shall be managed as a unit
or in close coordination.

4. Conservation and management measures shall not discriminate between resi-
dents of different States. If it becomes necessary to allocate or assign fishing
privileges among various United States fishermen, such allocation shall be (A)
fair and equitable to all such fishermen; (B) reasonably calculated to promote
conservation; and (C) carried out in such manner that no particular individual,
corporation, or other entity acquires an excessive share of such privileges.

5. Conservation and management measures shall, where practicable, consider efficiency in the utilization of fishery resources; except that no such measure shall
have economic allocation as its sole purpose.

6. Conservation and management measures shall take into account and allow for
variations among, and contingencies in, fisheries, fishery resources, and catches.

7. Conservation and management measures shall, where practicable, minimize
costs and avoid unnecessary duplication.

8. Conservation and management measures shall, consistent with the conservation requirements of this Act (including the prevention of overfishing and rebuilding of overfished stocks), take into account the importance of fishery resources
to fishing communities in order to (A) provide for the sustained participation of
such communities, and (B) to the extent practicable, minimize adverse economic impacts on such communities.

9. Conservation and management measure shall, to the extent practicable, (A)
minimize bycatch and (B) to the extent bycatch cannot be avoided, minimize the
mortality of such bycatch.

10. Conservation and management measures shall, to the extent practicable, promote the safety of human life at sea.

habitat and consideration of actions to conserve such habitat (section 110). Congress also required NMFS to convene a panel to consider ecosystem-based approaches to U.S. fishery management (NMFS in press).

Conservative Single-Species Management

The most obvious management approach is to reduce the catch of depleted species on a single-species basis. If Georges Bank is a prime example of the effects of overfishing in the United States, the case of striped bass on the U.S. east coast is a shining example of the effects of catch controls on a single species. Striped bass populations collapsed throughout the mid-Atlantic region and elsewhere in the late 1970s (Richkus et al. 1992). Chesapeake Bay is thought to be the nursery ground for 60 to 80 percent of striped bass off the east coast of the United States. In 1984, Congress passed the Striped Bass Conservation Act, giving states authority to place moratoria on fishing for striped bass. In 1985, Goodyear published calculations showing that control of fishing for bass would lead to a rebuilding of the populations, even if the decline had causes other than overfishing (Goodyear 1985). Led by Maryland, which imposed a moratorium on striped-bass fishing in Chesapeake Bay in 1985, and Virginia in 1988, the east-coast states increasingly controlled fishing effort. In early 1995, striped bass were declared by the Atlantic States Marine Fisheries Commission to be fully recovered (NMFS 1996a).

Other species have also responded to controls applied on a single-species basis. Indeed, Myers et al. (1995) concluded, based on an examination of life-histories, that almost all overexploited fish populations would recover if fishing were stopped. For example, both king (*Scomberomorus cavalla*) and Spanish (*S. maculatus*) mackerel catches off the southeastern and Gulf coasts of the United States have been severely restricted since the mid-1980s. Spanish mackerel were removed from overfished status to fully exploited status in 1995, and their populations have shown considerable increases (NMFS 1996a). There is some optimism that king mackerel populations will increase as well. Pacific halibut have long been managed on a single-species basis and have supported a sustainable fishery since the 1920s.

Several specific methods of implementing conservative management have been described. Marine protected areas are discussed in detail below. Another approach is to adopt a fixed exploitation rate (as opposed to a fixed catch) (NRC 1996b, Walters and Parma 1996). Another is to allow fish to spawn at least once before they are fished (Myers and Mertz 1998). Myers and Mertz pointed out that this approach was recommended more than 100 years ago (Holt 1895), but that other approaches, such as maximizing yield from somatic growth, had reduced its influence on management. They also provided practical guidance, emphasizing that susceptibility of populations to overfishing is very sensitive to the age at which they are first caught. Populations that can be caught while young but become sexually mature when much older (e.g., bluefin tuna, some cod populations) are particularly vulnerable to overfishing.

This approach is not always as successful, as described above. For example, Pacific ocean perch, severely depleted by fishing in the 1960s, supported almost no directed fisheries in the 1970s and 1980s. Their stocks were considered to be rebuilt only in the mid-1990s (NMFS 1996b, North Pacific Fishery Management Council 1997). This is not a complete surprise, as Pacific ocean perch are very long lived, but, even so, 30 years is a long time to wait for positive results. Pacific sardine populations declined drastically off the U.S. west coast in the early 1950s and were unmeasurably low by the 1970s despite essentially zero landings from about 1960. Only after the late 1980s did their populations begin to recover, and they are still low (NMFS 1996a). However, as described in Chapter 3, many small pelagic marine species like sardines and anchovies are subject to large, environmentally influenced fluctuations, so cause-effect relationships are not clear in this case. In general, a large reduction of fishing effort is a *biologically* effective method of conserving or rebuilding many marine fish populations, however disruptive it might be socioeconomically.

Although it is often effective, a conservative single-species approach alone is probably insufficient to sustain fisheries or ecosystems at acceptable levels of productivity. One reason is that it, like many other approaches, is difficult to implement and enforce. A considerable amount of political energy was needed to implement moratoria on striped-bass fishing; the International Pacific Halibut Commission, which manages Pacific halibut, was established by international treaty. More important, however, is that continued adverse ecosystem effects can accrue even when the target species is not depleted (see Chapter 3).

Although single-species management can be effective for maintaining population levels of individual species (e.g., Bering Sea groundfish and striped bass in the Chesapeake Bay), other organisms in the ecosystems may be affected through bycatch and trophic interactions. For example, current fisheries in the Bering Sea apparently are stable under single-species management, although earlier fisheries coupled with changes in atmospheric and oceanic circulation patterns probably contributed to declines in marine mammals and birds (NRC 1996a). Nonetheless, universal application of conservative management on a single-species basis would go a long way toward reducing overexploitation of the world's marine fisheries.

Reducing Bycatch and Discards

Reducing bycatch and discards is clearly a high priority for management and has been made a specific goal in recent national policies and international agreements. The matter has been addressed recently by the U.S. Congress in the revised Magnuson-Stevens Fishery Conservation and Management Act (see pp 78-80). The National Marine Fisheries Service has drafted a national bycatch plan (NMFS 1998). These and other efforts appear to have produced results.

Data for the years 1994 and 1995 suggest that bycatch and discard rates have declined since the mid-1980s as a result of several factors (FAO 1997d, Natural Resources Consultants 1998), including a decline in fishing effort for some important species, time and area closures, adoption of more selective fishing technologies, enforcement of prohibitions of discarding by some countries, and more progressive attitudes among fishery managers, users, and society at large with respect to problems resulting from discards. In addition, discards (but not bycatch) have been reduced by new technologies for using a variety of marine species and a greater use of many species for human consumption and for feed for aquaculture and livestock. All these efforts have reduced discards by several million metric tons since 1990 (FAO 1997d, National Resources Consultants 1998), and they have reduced bycatch as well.

Perhaps the most important overall approach is to stop treating bycatch as if it were a side effect of directed fishing. Instead, as proposed by Davis (1996), for example, the existence of bycatch should be recognized and dealt with in fishery-management plans as part of an overall exploitation of the marine community. Thus, catch quotas would be established for various gear types that reflect the mix of species those gears typically catch. Total fish removals would be accounted for if the catch quotas were based on the assessment of the species mix as a whole. Obviously, the size of the catch quota should be based as much as possible on information on interactions among the species involved (i.e., an ecosystem consideration). Under Davis's proposal, there would be no target catch or bycatch for each species; instead, there would be a total catch for groups of species. In Alaska, bycatch of halibut and groundfish is considered in setting and monitoring annual quotas, and the fisheries are closed when annual catch or bycatch quotas for individual species are reached (Pennoyer 1997).

Related to the multispecies approach to bycatch is the idea of individual bycatch quotas as opposed to fleetwide quotas or total catch quotas of bycatch (Alverson et al. 1994). The idea is that each individual fisher would be given an incentive to reduce unwanted bycatch, instead of everyone racing to catch their quota of target species before others in the fishery. This can happen even with individual quotas for the *target* species, because fishers want to avoid the restrictions imposed by bycatch limits on those who have not yet taken their quota of target species. One result of this is that the quota for the target species is not reached before the bycatch quota stops the fishery. However, establishment of individual quotas for bycatch as well as for the target species appears to allow better control of results (Trumble 1996): managers then have the option of keeping the target catch constant and reducing bycatch or of increasing the target catch while keeping bycatch constant. A program of individual vessel fishing quotas for halibut in the sablefish fishery in Alaska appears to have reduced discard mortality of halibut to about 136 t in 1995 as compared with 650 t in 1995 (Pennoyer 1997). In that program, the bycaught fish must be retained and landed if they are of legal size. Bycatch management and enforcement often require the use of observers on fishing vessels and it is often time-consuming and costly, even for some programs that hold individual vessels responsible for their bycatch (Pennoyer 1997).

In some cases, however, the above approach would not work. There is no rational way, for example, to develop a total catch quota for endangered species such as sea turtles caught in shrimp trawls. Even when the above multispecies approach is appropriate, there can be unwanted bycatch (i.e., bycatch of kinds and sizes of animals that cannot be used). In those cases three basic approaches to bycatch reduction have shown promise: changes in the pattern or intensity of effort, changes in the fishing gear used, and bycatch-reduction devices (BRDs) (Alverson et al. 1994, FAO 1997d, Natural Resources Consultants 1998). For example, the National Research Council (NRC) recommended a combination of two of the above approaches to protect endangered sea turtles (NRC 1990). To change the pattern and intensity of effort, the NRC recommended avoidance of certain sensitive areas at certain times and reduction of tow times. The NRC also endorsed the National Marine Fisheries Service regulations requiring the use of turtle-excluder devices, which are a form of BRD. Changing fishing gear could mean changing emphasis, for example from long-lines to trawls, or vice versa, depending on the nature and extent of the bycatch. It also includes changing details of the gear, such as mesh sizes and shapes (Bublitz 1996, Kennelly and Broadhurst 1996). Investigations of so-called active BRDs also show some promise (e.g., Loverich 1996). However, those devices, which are controlled by an operator in response to observations of catches, depend to a large degree on differences in fish behavior (as other BRDs do), and have not been easy to develop to date.

In assessing the effectiveness of approaches to reduce bycatch, it is important to consider whether the approach will result in an increase in fishing effort at other times and places, with perhaps adverse results. For example, Pereyra (1996) warned that, although midwater trawls for walleye pollock in the Bering Sea have extremely low bycatch (Alverson et al. 1994), they tend to catch smaller fish than bottom trawls, which have higher bycatch rates. Thus, requirements to use midwater trawls could adversely affect pollock populations and increase discards of small fish. This warning again can be seen as advice to consider as many aspects of the marine ecosystem as possible in developing fishery-management plans.

Discards, although related to bycatch, introduce additional difficulties because they often are unreported and sometimes are illegal. We have described the analysis of Myers et al. (1997), who implicated unreported discards in erroneous estimates of fishing mortality on northern cod (Chapter 2). Perhaps a compromise is necessary between the desire to prevent discards and the desire for accurate information on the effects of fishing. This, like the development of most fishery-management activities, is an area that requires cooperation among several groups of stakeholders (i.e., managers, scientists, users, fishers, processors, and so on). A promising development is research by Lowry et al. (1996) on the fate of gadoid fish escaping from the cod ends of trawls that indicates fairly high mortality for small fish. Larger fish seem to have much higher survival rates. This kind of information is enormously valuable for understanding currently unobserved fishing mortality, and more such research is needed.

Another area needing knowledge is the ecosystem effects of retaining bycatch as opposed to discarding it. To the degree that discarded bycatch consists of dead animals, the question is how much the dead animals contribute to ecosystem structure and functioning. As an example, Alverson has reported that some catches counted as discards by Alverson et al. (1994) are now being retained and used on land far from the place of capture (D. L. Alverson, Natural Resources Consultants, personal communication, 1997). How does that difference in disposition affect the marine ecosystem where the animals were caught? Is it perhaps better for the marine ecosystem to discard some kinds of bycatch than to use it on land? If so, what kinds of bycatch should be discarded and in what circumstances? The answers to those questions are known poorly, if at all, but are important to an understanding and intelligent management of the ecosystem effects of fishing.

Finally, we note that some of the options discussed elsewhere in this report, in particular mariculture and marine protected areas, might achieve socioeconomic and other ecological goals in addition to reducing bycatch by reducing effort.

Marine Protected Areas

For the purposes of this report, a marine protected area (MPA) is defined as a spatially defined area in which all populations are free of exploitation. A primary purpose of such "no-take" zones has been to protect target species from exploitation and to allow their populations to recover. Such protection has been shown to result quickly in increases in the number or size of individuals of many target species (see Table 5-1). MPAs can also protect critical habitats (like spawning grounds or nursery beds), provide some protection from pollution, protect the marine landscape from degradation caused by destructive fishing practices, provide an important opportunity to learn about marine ecosystems and species dynamics, and protect all components of a marine community (Agardy 1994, Allison et al. 1998, Bohnsack 1998). Protection against management uncertainty is another critical function of MPAs: the populations inside such areas can serve as a "bank" against fluctuations in outside populations caused by fishery-management difficulties or miscalculations. Finally, and perhaps most important, MPAs represent an opportunity to protect ecosystems.

Even small MPAs can result in rapid changes in local populations of fished species (Box 5-2, Table 5-1). Density and average size of fished populations often increase after protection. Even unexploited species can increase because of habitat protection (Russ and Alcala 1989). Larger individuals tend to have much increased reproductive output, suggesting that overall reproduction of a particular species may increase significantly after establishment of protected areas.

Despite the overall success of MPAs that have been established and studied to date, there are important limitations to their effectiveness and huge gaps in our knowledge about how they function within broader marine ecosystems. Protected areas do not always result in higher density of target species or in higher biodiversity (Ruckelshaus and Hays 1997). This may be because larger individuals exert predation pressure that limits the number of smaller prey species or even juveniles of their own species. It might also result because the ecosystem or community being protected has been so changed by human activity that its original condition is unknown. For example, Jackson (1997) described overexploitation of large herbivores on coral reefs (manatees, the now-extinct Caribbean monk seal, and sea turtles) that devastated their populations and fundamentally changed the ecosystems in many parts of the Caribbean by about 1800. On some reefs, subsistence fishing eliminated most large fishes as well.

In many cases it is difficult to demonstrate the effectiveness of protected areas because of a lack of baseline data. For example, of the citations in Table 5-1, only about half of the studies compared target species before and after reserve establishment. The rest compared areas inside and outside reserves, which does not adequately control for differences attributable simply to habitat quality. The overwhelming differences between some protected and nonprotected areas may make such considerations minor (e.g., Alcala 1988), but well-planned studies of protected areas are required for the full range of protective effects to be understood. MPAs are also less likely to be useful for species with highly mobile life-history stages like pelagic fish or planktonic organisms. In some cases, protected areas could focus on spawning grounds (e.g., for cod or whales), nursery grounds for young of various species, or migratory corridors. However, in other cases the spatially explicit definition of MPAs may not be biologically meaningful, and other management tactics might be better (e.g., temporary full closure of the fishery on the migratory striped bass in the eastern United States).

The large gaps in our knowledge about protected areas should be addressed. Improved understanding will enable more effective MPA design, management, and evaluation. The stability of organisms within them is of primary concern. Little work has been done on this topic, probably because many MPAs are recent. Despite the gaps in our understanding, enough is known to recommend substantial increases in the number and area of reserves (Allison et al. 1998).

The following kinds of information could help make MPAs a more effective tool. Much work is needed to better inform decision makers about the details of establishing and managing marine reserves. Much of the large body of theory and practical experience pertaining to terrestrial reserves (e.g., Meffe and Carroll 1994) cannot be applied directly to marine reserves (Allison et al. 1998) because of their openness (i.e., many marine organisms, especially juveniles, are carried large distances by ocean currents and thus can enter and leave areas unpredictably) and because marine ecosystems respond to environmental variations differently than terrestrial ecosystems do (see Chapter 3). In addition, the impact of protected areas on the full set of species in the area is generally unknown. Variation in the responses of different species with different life histories is also unknown.

A critical area of ignorance is how a protected area functions within the broader marine ecosystem of which it is a part and if protected areas export biomass or eggs and larvae into the surrounding communities. The export problem is particularly acute because the value of marine reserves as spawning banks depends on the movement of eggs, larvae, or juveniles out of the protected area. Understanding the relationship of reserve size and placement to this export function is a critical step in understanding the value of reserves (Allison et al. 1998). To date, some indications are that reserves can function as net exporters of juveniles or larvae and that population densities adjacent to reserves can be higher than in areas far from reserves (Russ and Alcala 1989), but these observations are scattered and preliminary. Other knowledge gaps include the impact of local and distant oceanic conditions on optimal reserve placement, the impact of natural patchiness of the environment on reserve function, the degree to which edge effects alter reserve-ecosystem functioning, whether populations of species with highly dispersive life histories can replenish themselves within reserves, the size required for reserves to support "natural" communities, and the impact of various nonexploitative recreational activities on reserve functioning.

Some Practical Considerations

Despite the clear advantages of marine reserves as management, conservation, and research tools, their effectiveness depends crucially on how well they are matched to managers' goals as well as to management outside their boundaries. For example, if a reserve is designed to protect an individual species, is enough known about ecosystem processes to predict the result with confidence (e.g., coral reefs as described by Jackson [1997])? Protecting an area can affect the dynamics of the ecosystem within it, and that can affect the abundances and dynamics of individual species in surprising ways. A study of the ecosystem described in Box 3-1 would lead one to expect that protecting sea otters would result in a decrease in kelp and a resulting decrease in organisms that depend on kelp for their habitat or food. This might be an undesired result. It is also important to know if the reserve is the right size and design to achieve its goals for the species or ecosystem of concern. To help increase the effectiveness of reserves, Allison et al. (1998) offered three design guidelines or questions that would help reserve designers assess the significance and trends of the threats to organisms in the proposed reserves.

• Will organisms inside the reserve be able to persist even if exploitation outside the reserve increases?

• Will organisms in the reserve be able to persist even in the face of episodic climate events (e.g., El Niños) and directional changes in climate?

• Will organisms in the reserve be able to persist despite increased pollution, species introductions, and disease?

Allison et al. (1998) pointed out that, although those questions probably could not be answered completely, the awareness of the threats would help designers to make the reserves more effective biologically.

In addition to the need for a clear identification and statement of the goals of any proposed reserve, it is essential to design them adaptively (i.e., in such a way that their effectiveness can be assessed scientifically). This implies the need for monitoring and for controls. Judging effectiveness is not easy (Allison et al. 1998), but it is possible and must be attempted. As an example, Polovina and Haight (in press) described a case in which comparison of a marine protected area with a control clearly showed that the MPA was effective in slowing the decline of spiny lobster populations in Hawaii. However, a large-scale environmental change appears to have led to their declines even in the reserve. The existence of a control and careful monitoring allowed learning to take place, and showed how important environmental fluctuations can be.

The final—and perhaps most important—practical consideration is that MPAs are not substitutes for fishery management, but are one of several tools in the toolbox. If all other methods of fishery management were abandoned, marine protected areas would have to be enormous to protect ecosystems and fisheries, especially to protect widely ranging species. In any case, all the normal (and difficult) management methods of working to involve stakeholders, enforcement, monitoring, and adaptive management need to be used.

The Scope of the Need for Marine Protected Areas

Currently, less than one-quarter of a percent of the sea is in areas termed marine parks, marine preserves, or no-fishing zones (McAllister 1996). The degree of protection in these areas is generally far less than that proposed here. There are very few marine areas in which all species and all aspects of the habitat are protected. How much of the marine environment should be included in marine protected areas for them to fulfill their primary functions of ameliorating environmental and management uncertainty, providing a source of eggs, larvae, and recruits to adjacent areas, and protecting critical habitat? An answer to this question is critically important but is one of the most difficult aspects of the emerging discussion on marine protected areas (Box 5-2).

Goals for the scope and purposes of MPAs need to be clearly and quickly articulated. The marine environment is under continued threats, and marine

TABLE 5-1 Some Examples of Protection of Fish by Marine Protected Areas

Area	Positive Effect?	Control	Size	Time Frame	Target Taxa	Reference
Kenya	Y***	Spatial			fish	Samoilys 1988
Philippines	Y***	Temporal		3 yr	fish	Alcala 1988
Philippines	Y*	Spatial/ temporal		3 yr	fish	Russ and Alcala 1989
South Africa	Y*	Spatial			fish	Buxton and Smale 1989
Florida Keys	Y**	Temporal		2 yr	fish	Clark et al 1989
Kenya reefs	Y***	Spatial			fish	McClanahan and Shafir 1990
Caribbean	Y**	Spatial	1 km	4 yr	fish	Polunin and Roberts 1993
Belize	Y**	Spatial	4 km	4 yr	fish	Polunin and Roberts 1993
					conch lobsters	
South Africa	Y***	Spatial/ temporal	46 km	2-5 yr	fish	Bennett and Attwood 1991
Florida Keys	Y***	Spatial	>100 km	20 yr	fish	Bohnsack 1982
Red Sea	Y***	Spatial		15 yr	fish	Roberts and Polunin 1993
Chile	Y*	Spatial/ temporal			snails	Duran and Castilla 1989
Kenya reefs	Y*	Spatial/ temporal			fish	McClanahan 1994

NOTE: *, **, and *** denote statistical significance at $p \leq .05, .01,$ and $.005$ respectively

biological resources continue to decline. Without a clear goal, it is impossible to generate the debate that expansion of MPAs requires or to begin designing and implementing protected areas before environmental damage makes that impossible. For those reasons, recent proposals to establish MPAs in 20 percent of the marine environment by the year 2020 provide a useful reference point for future consideration. The proposals also emphasize the importance of acting immediately to greatly expand the amount of area protected.

This number—20 percent—appears alarmingly and impossibly large at first but is based on a number of independent lines of argument that converge on the need for this general magnitude of commitment (Bohnsack 1994, 1996). Current understanding of marine ecosystems and populations cannot rigorously defend this number against all criticism, but it does provide a rationale for adopting a marine reserve program of this magnitude.

First, the current allocation of 0.25 percent of marine areas to reserves is too low to have an appreciable effect on marine populations across their range or

BOX 5-2
Marine Protected Areas in the Philippine Islands

A series of studies in the Philippines illustrate several important successes and challenges related to MPAs (Russ and Alcala 1989, 1994, 1996; Vincent 1997a, 1997b; Russ 1989; Pajaro et al. 1997). The studies focused on small (less than 100 ha) reserves at Sumilon Island, Apo, and one near Handumon village, Northwestern Bohol in the central Philippines.

In all cases the reserves were effective to varying degrees in increasing the abundance and diversity of fishes, and at Sumilon fish abundance increased in adjacent waters as well. The increases in the reserves were substantial in some cases; for example, biomass increased from approximately 1.5 to 18 kg per 1,000 m^2 in nine years of protection at Sumilon and from 1 to 10.5 kg per 1,000 m^2 at Apo (Russ and Alcala 1996). At Sumilon, the resumption of unregulated fishing reversed the gains (Russ and Alcala 1994, 1996). Data on Handumon are not yet available but the fishers perceive an increase in the seahorse populations there (Vincent 1997a; Amanda Vincent, McGill University, personal communication, 1997).

In all cases the involvement and support of local fishers were a prerequisite for any success of the reserves; similar findings were described by Dye et al. (1994) and Odendaal et al. (1994) in South Africa and in Chile. Castilla and Fernandez (1998) also presented arguments that successful management of small-scale fisheries in general requires this involvement of fishers. In all of the above cases, enforcement was a problem that could be solved only when local fishers were sufficiently committed to the reserves and sufficiently concerned about threats to their resources that they were willing to act together to enforce the rules and prevent poaching.

These cases demonstrate that even local fisheries using little modern technology can devastate local marine ecosystems. In some of the cases, most notably that of the seahorses under study in Handumon (Vincent 1996, 1997a, 1997b; Pajaro et al. 1997), much of the fishing pressure resulted from the large international market in seahorses outside the area. A total of 32 nations are involved in trading seahorses. In particular, China and Hong Kong have large markets; in the early 1990s, annual consumption of seahorses in China, Taiwan, and Hong Kong exceeded 41 t per year, or more than 14 million animals (Vincent 1996). The animals are prized as aphrodisiacs, antiarthritic agents, and as anticholesterol therapy (Vincent 1996).

In all the cases described, success required at least some knowledge of the organisms' biology, and in all cases, additional information about that biology as well as the animals' physical and biological environments was considered likely to improve the reserve's success. The seahorse reserve at Handumon could be successful despite its small size because seahorses do not disperse widely; the reserves at Apo and Sumilon were able to increase fish abundance because the adults of those species have small home ranges. Although the larvae of many coral reef fishes disperse long distances, protection allowed arriving juveniles to survive and grow (Russ and Alcala 1996). Over the long term, it probably will be necessary to protect larger areas if all the surrounding areas are overfished. It is equally true that any attempt to establish such MPAs without adequate knowledge of local socioeconomic conditions would surely fail.

marine ecosystems, and thus this level of reserve commitment does not meet current goals for marine reserves as a management tool. In terrestrial systems, slightly more than 5 percent of the earth's land area enjoys some sort of protected status (Groombridge 1992). Clearly, even that amount of protection, while help-ful, is not enough to prevent continuing loss of biodiversity. Marine systems are much more open, with more geographic exchange, than most terrestrial systems and thus need an even greater area of protection than terrestrial systems when species with dispersive life stages are involved.

A second line of evidence independent of the first is that goals of fishery management often focus on the protection of a certain fraction of the spawning stock (Clark 1996). If only 2 percent of the standing stock of a species is allowed to spawn, each individual must produce 50 offspring for the population to main-tain itself. Such high reproduction per individual is very sensitive to environ-mental conditions and can lead to the collapse of the standing stock. As a result, fishery managers often try to protect at least 20 percent of the population. Spawn-ers must then produce at least five offspring each, but this value is likely to be more easily achieved on a long-term basis than the 50-offspring requirement discussed above. An example of this approach was given by Bohnsack (1994), who calculated that protecting 20 percent of the habitat of red snappers (*Lutjanus campechanus*) would increase the total productivity of the fish population. That increased productivity would more than compensate the fishers excluded from the closed areas. These arguments are supported by information that fishing drastically reduced the numbers of spawners in several species (Mace and Sissenwine 1993, Goodyear and Phares 1990).

A third line of reasoning is an application of the inverse-square law. One of the goals of protected areas is for them to export eggs, larvae, or juveniles to other areas. The sea is a powerful dilution agent, and eggs and larvae of even coastal species will decrease in density as they spread out from a center of production. Settlement away from a protected area will decline rapidly with distance, and unless the protected area is very large, or occurs as a string of protected areas along a coast, the export of larvae and juveniles will be limited. Although research on this topic is critically needed, the large dilution capacity of the oceans suggests that a substantial fraction of habitats need to be larval exporters for reproductive individuals within reserves to have an effect outside reserves.

An Economic Argument for Marine Protected Areas

The establishment of MPAs has the potential to affect many fishers, espe-cially to the degree that they lack mobility and the MPA excludes them from traditional fishing areas. It also might appear to impose a heavy cost on industry with no offsetting benefits. Nonetheless, there are economic considerations that can actually favor MPAs. This report has emphasized the unavoidable uncertain-ties involved in resource management, great enough to lead to the collapse of resources in some cases (e.g., northern cod). MPAs could act as a hedge against such uncertainty if they are big enough and are properly designed. The expected economic return from exploiting the resource might be reduced by the MPA. If it is reduced, one can argue that there is a compensating tradeoff in the form of reduced risk: one gains protection against catastrophe.

The principle of hedging one's bets is widely followed in many economic activities where irreducible uncertainty is encountered. For example, investors in stocks and bonds who are risk averse are usually advised to diversify their portfolios of financial assets. Part of the diverse portfolio includes low-yielding, liquid, and safe assets such as U.S. Treasury bills. Although the overall yield from the portfolio is reduced in certain economic environments by this approach, risk is reduced as well. MPAs can be seen as playing a role like that of liquid assets in a financial portfolio. Reduced expectation of economic returns is offset by protection against future economic disaster.[1]

Institutional Structures

Much has been written about the need for management to better conform to the time and space scales of fisheries and fishing communities. Regional and global restructuring of political systems over the past few centuries has diminished, dismantled, and distorted local systems of decision making and authority, as Johannes (1977, 1978b) and others have shown for Oceania. There are often mismatches between natural and political boundaries, with either too many or too few political boundaries, contributing to risk-prone management. At one extreme, ecosystem approaches to fishery management are hindered by *too many boundaries,* the situation in which watersheds or other ecosystems are divided among the jurisdictions of multiple governments and agencies. Such problems require institutional change toward more effective regional management systems (NRC 1996b). The regional fishery management councils in the United States were formed to address the problem of jurisdictional mismatches and multiple boundaries, but the problem remains both within that system and in relationships between it and state and interstate management bodies.

However, at the other extreme, *erasure of boundaries* can diminish the capability of communities, tribes, and other local governing entities to impose controls on what is caught from and done to the local land and sea (Cronon 1983). Accordingly, one of the major institutional challenges to using ecosystem approaches is the construction, or reconstruction, of appropriate boundaries, tailored to each specific fishery and fishery-management issues. Management centralized in national or regional authorities must be balanced with involvement of

---

[1]The National Research Council's Ocean Studies Board has recently begun a more detailed study on these topics, entitled "The Evaluation, Design, and Monitoring of Marine Reserves and Protected Areas in the U.S." The study will examine the utility of reserves for fisheries management as well as for protecting ecosystems and biodiversity, with a report scheduled for release in 2000.

local stakeholders, communities, businesses, and property owners (Lee 1993, Pinkerton and Weinstein 1995, NRC 1996b, McCay and Jentoft 1996, Hanna 1998). A major challenge is developing institutional structures with sufficient complexity in scope and scale to be appropriate for complex and dynamic ecological systems (Ostrom in press). The boundary problem is political and must be addressed to develop effective fishery management.

International Developments in Fishery Management

Fisheries that cross jurisdictions and even national boundaries are problematic. The United States *makes* policy pertinent to its own fisheries; it tries to *influence* policies with respect to international fisheries. Those activities need to be viewed as separate efforts. Therefore, we consider the likely usefulness of international agreements, treaties, and conventions here.

International concern about the sustainability of global fisheries has resulted in the inclusion of precautionary approaches in several recent international agreements related to fisheries (e.g., Box 5-3). Three agreements form the basis for international fishery management and provide goals for national fishery management systems.

1. *The United Nations Convention on the Law of the Sea* (UNCLOS)—This treaty entered into force on November 16, 1994, and is one of the primary instruments for the sustainable use and development of the ocean and its resources. The convention is based on a philosophy of rational use conforming with environmentally sound development. Among its many features, the convention promotes the goal of sustainability of fisheries (Articles 61.2 and 119.1). Article 61.2 requires countries to ensure (through proper conservation and management measures) that living resources in their EEZ are not endangered by overexploitation. Of the top 20 fishing countries, four have not signed the treaty: Peru, South Korea, Taiwan, and the United States, although the United States does comply with the treaty.

2. *Agreement for the Implementation of the Provisions of the United Nations Convention of 10 December 1982 Relating to the Conservation and Management of Straddling Fish Stocks and Highly Migratory Fish Stocks* (General Assembly of the United Nations 1995)—This agreement added to the UNCLOS framework provisions for international management of highly migratory fish species (e.g., tunas, billfishes) and fish stocks that cross international borders. It was adopted on December 4, 1995, but had not been ratified as of June 1998.

3. *FAO Code of Conduct for Responsible Fisheries* (FAO 1995c and d)— This agreement (Box 5-3) is based on Agenda 21 of the Rio Conference on Environment and Development and is notable because of its focus on the precautionary approach and because it places the burden of proof that uncertainty allows an increased catch on the fishers, not only on the managers.

These three agreements commit signatory countries to sustain their national fisheries, cooperate to sustain international fisheries, address the problems of overcapacity and bycatch, base management on sound scientific information, and conduct many other activities that should improve fisheries in the future. Because these agreements are relatively new, it is difficult to gauge their impact. Like all international agreements, their merit will depend on how many major fishing countries ratify the agreements, whether signatory countries abide by the agreements, and the effectiveness of the limited enforcement provisions. Adoption of these agreements will promote the ecosystem-based approaches recommended by this committee.

Box 5-3 Code of Conduct for Responsible Fisheries

1. PRECAUTIONARY APPROACH AND BURDEN OF PROOF

12. Within the framework outlined in Article 15 of the UNCED Rio Declaration, the precautionary approach to fisheries recognizes that fisheries systems are slowly reversible, poorly controllable, not well understood, and subject to changing human values.

13. The precautionary approach involves the application of prudent foresight. Taking account of the uncertainties in fisheries systems and the need to take action with incomplete knowledge, it requires, *inter alia*:

    a. consideration of the needs of future generations and avoidance of changes that are not potentially reversible;

    b. prior identification of undesirable outcomes and of measures that will avoid them or correct them promptly;

    c. that any necessary corrective measures are initiated without delay, and that they should achieve their purpose promptly, on a time scale not exceeding two or three decades;

    d. that where the likely impact of resource use is uncertain, priority should be given to conserving the productive capacity of the resource;

    e. that harvesting and processing capacity should be commensurate with estimated sustainable levels of resource, and that increases in capacity should be further constrained when resource productivity is highly uncertain;

    f. all fishing activities must have prior management authorization and be subject to periodic review;

    g. an established legal and institutional framework for fishery management, within which management plans that implement the above points are instituted for each fishery, and

    h. appropriate placement of the burden of proof by adhering to the requirements above.

Composition of Management Institutions

Fishery management is primarily a social process. To make it more successful, goals such as ecosystem resiliency (Arrow et al. 1995) should be more explicitly linked to social values (Norton 1995), and to do that, it is necessary to create institutions and organizations that help involve stakeholders in the process. The mobilization of society to reject unacceptable risks and to lobby for more conservative resource use may be a significant means of confronting uncertainty in future resource conservation and management and of developing risk-averse management (Alverson et al. 1994). A similar view was expressed by Parrish (1995), who noted that in many areas of the world one can determine when stocks are overfished and depleted, but few countries have political and fishery management systems capable of managing stocks at a fishing mortality rate that even approaches optimal levels.

Stakeholder involvement in fisheries and watershed management can be achieved through a range of mechanisms, from public hearings and advisory groups to collaborative or partnership management by governmental or nongovernmental bodies to community-based management (Jentoft and McCay 1995, McCay 1995b, Pinkerton and Weinstein 1995, Hanna 1998). Butterworth et al. (1993) described a process where industry representatives—the users—were included in setting catch rates during the fishing season for the South African anchovy (*Engraulis capensis*). There is a great deal of uncertainty as to the stock size in that species, with a midseason survey being required to set reliable catch quotas. The industry involvement provided a more effective way of dealing with uncertainty than simply having the management agency alone set catch quotas. A somewhat similar process is used in setting catch quotas in the United States through the regional fishery management councils.

How should fishers and other individuals who gain economic benefits from natural resources be involved in management processes? A risk of participatory democracy in fishery management that must be avoided is the potential for the process to be captured by narrow interests (Jentoft and McCay 1995, McCay and Jentoft 1996). Several studies of the U.S. fishery-management system under the Magnuson-Stevens Fishery Conservation and Management Act have highlighted conflicts of interest as a major cause of regional council failure to prevent overfishing (Ludwig et al. 1993, WWF 1995). In addition, the workings of the regional councils tend to reflect and perhaps increase competition among fishing sectors and engender adversarial relationships with the National Marine Fisheries Service, as well as increased reliance on lawsuits and congressional involvement. These and other problems somewhat reduce the benefits of participatory and regionalized decision making.

On the other hand, early and meaningful involvement of members of the public can have important benefits (Hanna 1995). The approach of involving stakeholders in the regional councils is an innovative one, and in some of the regions the process works better than in others, leading to optimism that the process can be improved. As another example, the NRC (1996b) has recommended the development of management institutions to conserve salmon in the Pacific Northwest that include a regional or watershed-scale component, that allow for shared decision making among all legitimate interests, and that ensure that local or regional interests not be permitted to override the interests of the greater region.

Sand (1992) found that failure to enunciate clear goals has constituted an institutional flaw in many environmental agreements. Miles (1994) suggested that institutional performance could be improved by

- matching fishery effort and resource availability (i.e., reducing overcapacity);
- broadening the focus of fishery management to include all sources of environmental degradation that affects fisheries;
- structuring the duty to cooperate and conserve through a set of international principles;
- implementing effective monitoring and enforcement principles; and
- creating institutions with the capacity to mandate collection and exchange of vital data.

SOCIOECONOMIC INCENTIVES

Much of fishery management, at least from an economic standpoint, is designed to cope with the problems created by the current economic incentive system described in Chapter 4. Two broad approaches—not mutually exclusive—have been tried: regulation of effort and a rights- or privilege-based approach.

Regulation of Effort

Regulation of effort is an attempt to prevent fishers from responding to economic incentive systems in ways that society deems unacceptable. Various controls are put into effect, for example, setting total allowable catches, imposing net-mesh regulations, limiting boat size, limiting gear type (e.g., restricting ocean salmon fishing to troll gear), seasonal restrictions, and controls on the number and kinds of vessels entering a fishery. These aim to prevent overexploitation of a fishery resource, in part by reducing excess fishing capacity.

The regulatory approach has unquestionably had some successes in protecting fishery resources. Examples include Pacific halibut, walleye pollock in Alaska, and various cases where the controls became absolute (i.e., the fishery was completely halted, such as striped bass in waters of the eastern United States). However, regulations have not in general reduced excess capacity. Fishing capacity represents many inputs, and it is almost impossible for resource managers to control all the inputs. As long as the economic incentives remain in force, fishers attempt to substitute unregulated inputs for regulated ones (Dupont 1996). As described in Chapter 4, fishers are ingenious in circumventing regulations to limit entry and fishing power. This recognition has led to the conclusion that the most effective way to reduce fishing power is to address the incentive problem. This means that one needs to identify a system that reduces or removes fishers' rewards for overcapitalization and that increases the likely future returns on their investments, so that investment by conservation or enhancement is in their interest. Two approaches to development of such systems are described below. Although at first the approaches appear to be polar opposites, they do have features in common and in some circumstances appear to have the potential to coalesce. The following discussion should be taken as providing examples of approaches to a problem that the committee believes must be addressed, rather than as an endorsement of any particular scheme over any other. Whatever system is used, it appears that including the users in the decision-making process is important to achieving successful outcomes. To ease the socioeconomic and political realities of changing from one regime to another, some process of "grandfathering" rights or privileges should be considered, as it often is when political or legal regimes change (e.g., tax laws, changes in Social Security).

The Opportunity to Participate in Fisheries

A key issue in the management of fisheries is who has the opportunity to fish. Historically, the opportunity to fish has been open to all (referred to as *open access*), and this has led to the many problems described in other sections of this report. Even when access has been controlled or limited, these problems are usually not corrected. The problems are a result of fishery-management systems that make participants in a fishery competitors for a share of the available resource, which makes it rational for them to try to quickly catch a larger share. While this behavior is rational for individuals, it drives up total costs, so that the net economic benefits from fisheries dwindle or may be negative (at least for a while), and participants do not invest in conservation for the future, largely because they have no assurance that they will be beneficiaries in the future. The essence of the problem is that a fishery-management system that makes participants compete for shares of the resource does not provide incentives for efficiency and conservation.

The alternative are fishery-management systems that assign exclusive rights or access privileges to a share of the resource. In this situation, the incentive is to use the shares efficiently rather than to spend more in competition for a bigger share. And if participants' rights extend into the future, there is an incentive to make conservation investments because there is some assurance of the opportunity to benefit in the future. This form of management is often referred to as *rights-based.*[2]

There are many forms of rights-based fishery-management systems, usually specified in terms of the definition of exclusivity, the rules about transferability of the right, and the nature of the right. This specification of the management system has a great deal of influence on the effectiveness of incentives for efficiency and conservation. Exclusivity may apply to individuals, corporations, government entities, communities, or other groups. For the specification to result in positive incentives, the exclusivity must be assigned to entities that are cohesive enough to act for the collective good of the entire entity (i.e., as quasi-individuals). Rules about transferability may range from disallowing it to allowing it without restrictions. Transferability increases economic efficiency by allowing rights or privileges to be transferred to entities that can use them most efficiently (i.e., with the lowest cost of fishing). But there are many reasons for restricting transferability, such as to prevent monopolies or for social reasons (e.g., to preserve the traditional nature of participation in fisheries). The nature of the right or privilege may be a specific amount of catch, an annual share of a total allowable catch, a specific amount of fishing effort or units of fishing capacity, or a geographically defined fishing area. Some types of specification may not completely eliminate the incentive to compete for a larger share of the resource, but they may have other positive attributes, such as lower enforcement costs or greater social acceptance. Box 5-4 describes a successful example of limiting access through permits and quotas.

The most appropriate specification of rights-based management will be a compromise between management objectives and constraints that varies from one case to another. But it is clear that some form of rights-based management that instills positive incentives for efficiency and conservation is needed.

Community-Based Management

A promising approach to rights-based management is to award rights to communities. These include place-based communities, like municipalities or the

---

[2]The following discussion covers rights, privileges, permits, and related methods of restricting access to a fishery. The differences among those forms of access limitation are important, but are beyond detailed discussion in this report. We note here that the Sustainable Fisheries Act of 1996 is clear in specifying that "any individual fishing quota or other limited access system authorization" is to be considered a permit (rather than a property right), does not confer any right of compensation (as would a property right under the 5th Amendment to the U.S. Constitution) if it is revoked or limited, and "shall not create, or be construed to create, any right, title, or interest in or to any fish before the fish is harvested" (Section 108 [e]). The act prohibits the establishment of new individual fishing-quota systems before October 1, 2000 and mandates two National Academy of Sciences/National Research Council studies—currently in progress—to evaluate the use of individual fishing quotas, community-development quotas, and other rights-based systems.

BOX 5-4
The Herring Fishery in San Francisco Bay

The fishery for the herring (Clupea harengus pallasi) in San Francisco Bay is an example of a fishery that appears to be maintained in a sustainable fashion by using traditional single-species management strategies. The fishery has been characterized by three major peaks in landings in response to new demands. The first peak occurred in 1918, when most of the fish were processed into fish meal. When the reduction of whole fish into fish meal was prohibited in 1918, the fishery ended. From 1947 to 1954, whole herring were harvested and canned to make up for declining sardine stocks. The fishery again declined. Since 1973, a new international market has developed for herring roe, a delicacy in Japan (Spratt 1981).

Pacific herring live most of their adult lives in the ocean and return to San Francisco Bay only to spawn. Large schools of herring enter the bay during spawning season and may remain for up to three weeks. After spawning, they return to the ocean, where they are planktivorous, feeding primarily on zooplankton. Other marine fish and birds may forage on herring, but no higher predator depends on a diet of herring alone. Herring mature at age two and may live for 10 years, making each fish capable of several spawning migrations into the bay (Ware 1985).

San Francisco Bay provides over 90 percent of the state's herring catch (Spratt 1981). The fish must be caught within one day of spawning or while spawning is in progress, a season from November to March. A secondary fishery exists for the roe deposited on kelp fronds. Giant kelp is removed from the vicinity of the California Channel Islands and suspended from rafts in San Francisco Bay in areas where spawning is likely. After spawning, the kelp is collected with the roe attached.

Management of the fishery is based on population estimates from annual hydroacoustic surveys and spawning ground surveys. Quotas are set at about 15 percent of the amount of herring expected to return to spawning areas. These quotas are adjusted annually, and the maximum catch rate is recommended to be 20 percent (Trumble and Humphries 1985). The estimated population of Pacific herring in San Francisco Bay declined in the 1983-1984 season, probably in response to the 1983 El Niño. However, the population has been rebuilding since 1984, and spawning biomass was approximately 65,000 tons from 1987 to 1990 (Spratt 1992).

The primary tool of management is limiting entry to the fishery. Since 1983, only five new permits have been issued, and the total number of permits in San Francisco Bay is stable at about 400 for fishing and 10 for the roe-on-kelp fishery (Spratt 1992). Regulations change yearly and respond to new conflicts that arise. Several new techniques have been used: permits issued by lottery, individual vessel quotas, and allowing the selling of permits. All of these are single-species techniques, yet they have provided effective management of the fishery.

fishers of a particular area; interest-based communities (i.e., groups of fishers, fish processors and others who work in the same fishery, including cooperatives and corporations); and a broader conception of community, here called *virtual communities*.

There are examples of coastal communities that succeeded in managing fishery resources sustainably and avoiding overexploitation (Christy 1982, Cordell 1989, McGoodwin 1990, Hviding and Baines 1994, Pinkerton and Weinstein 1995, Leal 1996), although in some cases, the demand for food can overwhelm the ability of coastal communities to manage their resources (Simenstad et al. 1978, Jackson 1997). How have the successful communities overcome such problems as ill-defined property rights, which contribute to the overcapitalization that seems so important to the overexploitation problem? Pinkerton and Weinstein (1995) identified some prerequisites for the success of such schemes, including a minimal degree of exclusivity with respect to the resource, a high degree of community dependence on the resource, and the ability to assert management rights on an informal, if not formal, basis.

Until recently the focus has been on place-based communities, like local tribes and cooperatives or fishing villages. There are promising new directions being taken, including the use of community quotas, such as the community-development quotas allocated to certain community organizations of western Alaska (Ginter 1995). With the development of interest in "comanagement" arrangements, where government bodies and groups of resource users share the rights and responsibilities of managing resources (Jentoft 1989, Pinkerton 1994), other forms of community have been recognized as well. For example, the Canadian government is entering into "partnership" agreements with groups of licensed fishers who establish their own management plans within frameworks established by the government. There has also been discussion of the prospects of a corporate model for fishery management (Townsend 1995, Townsend and Pooley 1995). Moreover, some advisory groups for government management agencies have begun to function as communities of collaboration among industry members, scientists, enforcement officers, recreational anglers, and others for fishery management (McCay et al. 1995). Based on these and other ideas discussed at the committee's conference in Monterey in February 1996, the committee recommends expanding the notion of community in fisheries management, using the term *virtual community*.

A virtual community is the functional equivalent of a geographically defined community in which the members of the community may not all reside in the same area, as noted by Reingold (1993) in his discussion of a similar phenomenon among Internet users. It may be thought of as a community of interest, as distinct from a community of place. Some communities of interest might be narrowly defined groups of fishers who share a common interest in a particular management regime and are granted some exclusivity with respect to the resource, or a share of the resource, in exchange for taking on responsibilities in managing it. Others might be made up of landowners, whose activities impinge on habitats important to fish species. Even more promising are virtual communities that constitute all parties sharing an interest in a fishery and its associated habitat. Such communities would include a very wide range—fishers and fish plant workers, landowners, biologists, community-development groups, recreational anglers, and conservationists—similar to watershed associations.

Like many small-scale, fishery-dependent coastal settlements, virtual communities have the potential to provide the social framework for managing fisheries sustainably. The essence of community is mutual communication, shared understanding for the need to solve problems, and collective action. In this broader concept of a community of interest and collaboration, information on the fishery and the larger ecosystem would be shared on a regular basis. As in virtual reality, where computers provide an opportunity to simulate what happens in the environment, a virtual community could provide computer-based modeling and data-sharing networks. In the Pacific Fishery Management Council's salmon-management process, some fishery leaders are already using fishery models to assist them in reaching consensus with fishery scientists on appropriate harvest levels.

Information would also be openly communicated about the diverse and sometimes conflicting values and needs of the human components of that ecosystem. Ideally, in such a community, trust and mutual knowledge would develop to enable those involved, however diverse their interests, to identify and work on common grounds and develop mutually acceptable goals and visions for the future of the resource and the ecosystem. Accordingly, such communities of mutual interest would come closer to realizing the objectives of ecological approaches to natural-resource management, which includes bringing user groups, local communities, and other members of the public into productive collaboration with scientists and managers (Kessler and Salwasser 1995). Experimental, game-theory, and comparative case-study research over the past decade have shown that communication, trust, and reciprocity are important requirements for groups to engage in the kind of collective action needed for community-based resource management (Ostrom 1998). The degree to which virtual communities and place-based communities have these and other conditions important to the exercise of stewardship also depends on the degree to which the activity in question—fishing in this case—is the primary reason for the community's existence.

Whether based on communities of place, interest, or collaboration, rights-based management is not necessarily dependent on exclusive rights of use or access. Use or access rights are not the only kinds of rights that make a difference in the ability of a community to manage resources sustainably. The broader conception of rights-based management that the committee advocates recognizes, first, that with rights come responsibilities, and, second, that among the critical rights are rights to information—to make policy, to plan, and to coordinate with other uses (Pinkerton 1997). It is then possible to develop more appropriate institutional arrangements for fisheries. For example, the virtual community may not hold exclusive access rights to a fishery but may establish the conditions and constraints for those who do hold access rights in ways that combine long-term stewardship interests with shorter-term economic imperatives. Contractual arrangements among those with communal rights, for some purposes, and those with individual rights, for other purposes, also can be envisioned (Rieser 1997).

Individual Transferable Quotas

A popular, albeit controversial, management scheme for fisheries designed to alter economic incentives involves individual catch quotas. Commonly the individual quotas are transferable, and hence they are known as ITQs (individual transferable quotas). ITQ schemes have been put in place for halibut in Canada and Alaska (Box 5-5) sablefish in Alaska, wreckfish (*Polyprion americanus*) in South Carolina in 1992 (Gauvin et al. 1993), surf clams (*Spisulas solidissima*) and ocean quahogs (*Arctica islandica*) off the northeastern United States (McCay et al. 1995), and various species in many other parts of the world, especially Australia, New Zealand, and Iceland since about 1985. The basic idea is that the

---

BOX 5-5
ITQs in British Columbia and Alaska Halibut Fisheries

Despite the existence of a limited-entry system in the Canadian fishery, the length of the season declined from 60 to 6 days within a decade. The industry approached the Canadian Department of Fisheries and Oceans (DFO) for help. DFO worked out an individual fishing-quota program for a two-year term in 1991 and 1992. Quota transfers were initially prohibited. The system was extended another two years with partial transferability, with no more than four shares per vessel. A modified system is still in operation, and quota owners consider the system to be a success. It is an improvement over previous systems from a conservation viewpoint because in the previous 10 years the quota had been overrun seven times and there was too much capacity in the fishery, resulting in significant waste. There are now higher exvessel prices (the price obtained by the fisher when the catch is sold to the wholesaler or seafood broker), fresh fish available throughout the year, and an increase in the number of buyers. Innovative marketing has been implemented. Users pay the monitoring cost. This is a good example of cooperative government-industry problem solving.

Similar difficulties affected the Alaska halibut fishery. The fishery's management had been successful in protecting the resource, but socially and economically it had unfortunate results. The fishery was characterized by overcapitalization and an extreme "derby" fishery (Buck 1995). As a result, an ITQ system was promulgated in 1995 (NMFS 1996a). The results have not been entirely to everyone's liking, but individual quotas have been well received by many segments of Alaska's halibut-fishing industry (Smith 1997).

authorities would set a total allowable catch (TAC), and then divide the TAC among individual fishers or groups of fishers (companies). The individuals, no longer having to compete for shares of the catch, would no longer be rewarded by overcapitalization. Described this way, ITQs—being *individual* transferable quotas—appear to represent a fundamentally different kind of ownership from virtual communities, in which ownership of the resource is collective. But since ITQs often develop into more than merely claims to catch shares, there is potential for ITQs and virtual communities to be complementary rather than conflicting (Scott 1993). Indeed, Alaska's community development quotas (CDQs)—catch quotas allocated to coastal communities— represent an evolution in this direction, and Castilla and Fernandez (1998) described how ITQs evolved in this fashion in an inshore fishery for invertebrates in Chile.

Combined Rights-Based Approaches

Although ITQs mitigate some adverse economic incentives, they do not necessarily reduce the discount rates experienced by fishers enough to ensure resource conservation (see, e.g., Gauvin et al. 1993, Mace 1993). That is because the productivity of the resource may be lower than the economically rational discount rate, a limitation of all forms of allocation. Furthermore, ITQs seem likely to work only if effectively monitored and enforced by authorities. However, in some cases, holders of ITQs may form virtual communities, by themselves or with others. These communities exercise collective responsibility for management of the resource (Scott 1993), as noted in New Zealand (Annala 1996), Iceland (Arnason 1996), and Nova Scotia (McCay et al. 1995), and thus can improve the enforcement of the well-defined rights conveyed by ITQs. In those countries (as elsewhere) the ITQs are expressed as percentages of the TAC rather than as fixed quantities. Thus, if the resource declines, the quotas' values decline as well, and that provides an incentive for investment in the resource (i.e., wise management). In this way, they take on some (but not all) of the attributes of shares in a corporation. They are claims to the stream of economic return from the natural capital (i.e., the resource), and their value will reflect the capitalized value of the expected returns from the resource. Expected consequences of management (and of natural fluctuations and environmental changes) will be reflected in the price of the ITQs, and thus sound management should be rewarded. Indeed, the price of wreckfish ITQ shares is stable and fish prices have increased (NMFS 1996b). To the degree that this occurs, the ITQ holders are taking on some of the characteristics of shareholders (i.e., they become de facto collective owners of the resource). An example is that of ITQ owners in a New Zealand abalone fishery who levy a charge on their sales to fund research and stock-enhancement programs of their association (Pearse and Walters 1992).

Clearly, a great deal more experience is needed with such schemes to understand their ramifications, the circumstances in which they might work well and in which they might fail, and the appropriateness of variations on the themes. Some of these questions and practical experiences with ITQs have been discussed by McCay and her coworkers (McCay 1995, McCay et al. 1995) and OECD (1997). Concerns about ITQs include questions about equity (e.g., the assignment to individuals of exclusive rights to exploit resources that are perceived to belong to everyone and whether the shares should be given away or sold), concentration of the rights or shares in very few people's hands, questions about their effectiveness in promoting stewardship, their effect in reducing the number of participants in fisheries, the best extent and duration of the rights or shares, and other related concerns. In addition to researching those concerns, there is room to consider broadening the scope of the virtual communities, especially if one views fisheries in an ecosystem context. Should timber interests be included in salmon-fishery virtual communities, for example?

Hanna (1998) provided an analysis of ways to adapt institutions and property-rights regimes to an ecosystem approach to fishery management that reflects attributes of the ecosystem and its human users, values ecosystem services, and coordinates interest groups and managers on a broad ecosystem scale (see also NRC 1996b). Management structures need to promote the definition of multiple objectives through processes that are legitimate and flexible and that promote socially appropriate time horizons for resource use and decision making. They need to take uncertainties into account, including what Hanna calls *fact uncertainty* (lack of knowledge) and *tenure uncertainty* (resulting from unspecified property rights or uncertainty in political systems). Despite these uncertainties, the recent developments of various rights-based allocation schemes offer considerable hope for sustainable fishery management. Indeed, it not clear what other general course offers as much promise.

Managing complex biological systems is difficult because of the often large differences between the social and temporal scales of natural and socio-political systems (NRC 1996a, 1996b). Although incentive structures created by allocating transferable use rights to private entities may promote greater efficiency and new ways of valuing the resources, unless they are designed correctly, they may not by themselves adequately protect and enhance ecosystem goods and services by protecting habitat, preventing pollution, and coordinating with other fisheries (Scott 1993). Hence, other forms of organization, including comanaging and virtual communities, are needed as well. In the broader sense of rights-based fishery management advanced in this report, different kinds of rights, ranging from rights to a resource to rights of governance, would be combined with different forms of ownership, ranging from individual ownership to ownership by communities and to ownership by the public or national government. The public-trust nature of marine resources as well as public rights in the ecosystem goods and services provided by marine environments (Rieser 1991) can be combined with private and community-based ownership of rights, access, capture, and management (Rieser 1997). Such institutional complexity is also necessary in managing the biological complexity of marine ecosystems (Ostrom in press).

The above discussion focuses on commercial fisheries, but recreational fishing is also central to the sustainability of many fisheries and is subject to economic incentives that can inhibit (or promote) conservation. An example of these incentives is provided by marine sportfishing tournaments. Although many such tournaments require release of the captured fish alive, many require at least some fish—usually dead—to be weighed to be eligible for the prize, which can exceed $100,000 in cash and equipment for the heaviest fish, with additional prizes for runners-up and for other categories. A recent (summer 1998) search of the World Wide Web turned up information on dozens of saltwater fishing tournaments with total purses of up to $1 million, and sportfishing magazines have many advertisements for such tournaments in every issue. There are also many smaller tournaments with much smaller purses, often for smaller species of fish such as bluefish and various mackerels.

For a 500-pound fish, a prize of $100,000 would amount to a price of $200 per pound, much more than the usual value of a fish caught for food. In addition, the prize money is uncertain: it depends on what others have caught as well as the willingness of the sponsor to pay. Therefore, such tournaments—which are popular throughout the coastal United States—can encourage the killing of fish for uses other than personal consumption or even sale for food. For large fishes at high trophic levels or for long-lived, slow growing species, such tournaments can contribute significantly to the overall fishing mortality. As the importance of such mortality has become clearer, many sponsors, especially of billfish tournaments, have moved toward tournaments in which prizes are given for fish that are released alive, and to the degree that such tournaments can be substituted for tournaments in which fish must be killed to earn prizes, an important economic incentive to kill fish would be eliminated.

The principles and objectives of rights-based management apply to recreational and subsistence fisheries as well as to commercial ones. For example, where there is private or association ownership of recreational fishing sites, as in many inland rivers and lakes, there may be substantial efforts to protect habitat, prevent pollution, and work toward enhancement of fish populations. A notable example is that of community-run systems for managing moose, salmon, lake trout, and other species in Quebec (Leal 1996). Moreover, if rights are allocated to commercial fishers, it is possible for groups of anglers to buy out those rights for their own purposes, including conservation; for example, the North Atlantic Salmon Fund bought rights of salmon fishers from Greenland and the Faeroe Islands beginning in 1991. Furthermore, all groups with well-defined rights are thereby in stronger positions to use the courts to protect fishery systems from pollution and habitat destruction (Brubaker 1996).

SCIENTIFIC MATTERS

Understanding Marine Ecosystems

The ability to use ecosystem approaches to sustain marine fisheries will depend on better information. Managers need understanding and models that encompass components of ecosystems (including humans) and information about whether they are changing and if so, how; the causes of change; and how negative changes or impacts might be reduced. An understanding of the importance of the impacts of local activities to much larger spatial and temporal scales is crucial. Ocean science can contribute the requisite information and must also be tapped to develop new tools for observing and managing fish populations and marine ecosystems. Fishery managers are required to use the "best scientific information available" (MSFCMA National Standard #2). Unfortunately, the best scientific information may not be communicated to policy makers and, even if communicated, is not always adequately used. Thus, this aspect of science related to implementing ecosystem approaches must be linked closely with institutional mechanisms that specify the information needed and communicate it to managers, policy makers, and the public.

The only way to anticipate how ecosystems will respond to perturbation is to develop a better understanding of how mechanisms at lower levels of organization provide the feedbacks that govern the dynamic and nonlinear features of ecosystem responses. Marine ecosystems are assembled from loosely coevolved species into assemblages. They have the capacity for multiple stable states and complex dynamics, including chaotic fluctuations. A static view that is restricted to a description of flows is inadequate to understand the responses of a system beyond the range of conditions it has previously experienced, as well as the potential for regime shifts or other qualitative responses to climate change or fishing pressure (Steele 1998).

Ecological systems are complex interconnected nonlinear systems; as such, their dynamics may be very sensitive to past conditions, and subject to shifts in dynamics when exposed to environmental stresses or sustained fishing pressure. Keystone species such as the sea otter (see Box 3-1) are important because of their potential to mediate such domain shifts, which can have dramatic consequences for marine fisheries. The collapse of the Barents Sea herring stock because of overfishing is a case in point; with the demise of the herring, pressure on capelin was reduced, leading to an increase in those stocks; on the other hand, cod populations declined owing to inadequate food supply, and whale populations changed their range, with important effects on the dynamics of other ecosystems.

The resilience of an ecosystem can be defined as its capability to maintain essential structure in the face of perturbation and to resist significant shifts in dynamics. This is not an adaptation of the system in an evolutionary sense, but it is an emergent property with profound importance to humans. Flexibility is the ecosystem-level equivalent of the ability of species to "adapt" in response to environmental changes in ways that lead to persistence. For the individual species that flexibility is embedded in its genetic diversity; the same idea applies at the ecosystem level, where biotic diversity is equally important to resilience. Management for sustainability means preservation of biodiversity both for its own sake and because of its importance in maintaining ecosystem resilience.

Marine ecosystems are often defined by geographic boundaries, such as the Bering Sea. They can be large and can overlap geographic and political boundaries; their own boundaries are often not easy to delineate or define (Alexander 1990). In this discussion, we use the concept of the large marine ecosystem described by Sherman (1990) as "relatively large regions of the world, generally on the order of 200,000 km^2 [or more], characterized by unique bathymetry, hydrography, and productivity. . ." The geographic scale of most marine ecosystems is larger than the scale of the local human communities that depend on them. Furthermore, marine ecosystems are frequently include discontinuous habitats with substantial inputs (e.g., nutrients, sediments, energy, invaders) from other kinds of habitats in the ecosystem or from other ecosystems. Because of this coupling of habitats in different areas, management schemes at the local level must include the effect of local decisions on the larger ecosystem and over long times.

Fishes and other components of marine food webs have complex life histories with different habitat requirements at each of several stages. Differences among species in spawning grounds and dispersal ability have important implications for management. Knowledge of habitat requirements and of timing of settlement of larval dispersal stages is needed to understand the effects of localized changes in environment and to predict the strength of interactions among species with complex life histories.

Understanding Policy, Institutions, and Behavior

Recent approaches to resource management place humans and their institutions squarely within ecosystems (Pickett and Ostfield 1995) and recognize the need for constructive and broad-based participation of the public in policy making and implementation (Kessler and Salwasser 1995). As experience with comanagement, virtual communities, participatory research, ITQs, and other institutional innovations increases, so does the importance of designing them appropriately. If it is important to match institutional scales of complexity with biological ones (Ostrom in press, NRC 1996a, 1996b), more work is needed to determine how this can be done in general and in particular cases. The development of comanagement and community-based management raises many issues, including ones of balancing interests of local and interest groups with those of larger publics and longer-term ecological systems and finding ways to gain the benefits of broad-based participation without sacrificing the benefits of small-group participation and highly informed input into policy making (McCay and Jentoft 1996). The focus on rights-based management and its potential for changing incentives to foster more stewardship raises questions about whether exclusive use rights are necessary to meet that goal or whether other kinds of rights and responsibilities, including those to manage resources or habitats, can work as well or better than exclusive rights to use or access (Pinkerton 1997). The promising new direction of ITQs has a host of related issues. Most central to the question of sustainable marine fisheries is the question of how and whether ITQ management regimes can be designed to realize the benefits of a market-based approach, in terms of efficiency, while also providing incentives for greater responsibility and stewardship (Young and McCay 1995, Brubaker 1996).

Social-science investigations are discussed in various parts of this report, and they need to include research on the structure and functioning of virtual communities, on the choices people make in the face of various management and economic regimes (e.g., ITQs), on people's and communities' adaptations to changing employment opportunities in fisheries, and so on. In addition, research into institutional structures and how they function is important (e.g., NRC 1996a, 1996b).

Data and Monitoring

Valid scientific recommendations are frequently ignored when fishery regulations (e.g., regarding quotas and seasons) are implemented. Collection of reliable data over long periods in the correct locations is crucial for developing better understanding of fish population dynamics and marine ecosystem function. Monitoring is important for detecting trends and patterns of variations over time so that causal relationships among biological and physical factors can be determined. Monitoring and assessments allow evaluation of existing and new management approaches and contribute information useful to research scientists. They are also necessary to gain a better understanding of human and natural effects on fish populations and marine ecosystems, so that predictive and diagnostic models can be created and may indicate new research topics that should be pursued. Analysis of information collected at regular intervals over extended periods should lead to better understanding of patterns of variation and linkages between human population growth, utilization of resources, environmental degradation, and climate change. The importance of such information is indicated by the development of industry-based data bases, such as the Groundfish Data Bank operated in Kodiak, Alaska, for fishers in the Gulf of Alaska and the Bering Sea.

Most existing monitoring and assessment programs consist of periodic sampling by government agencies (states and the National Marine Fisheries Service in the United States; the Department of Fisheries and Oceans in Canada). There are also government-funded programs carried out to monitor fishery ecosystems for the long-term goal of understanding system dynamics. This type of monitoring is exemplified by observations of the California Current ecosystem conducted by the California Cooperative Oceanic Fisheries Investigation (CalCOFI). This program has accumulated one of the most extensive long-term data sets of ecosystem factors related to fisheries. Other similar programs include various activities of the International Council for Exploration of the Sea and such activities as the Continuous Plankton Recorder Survey, which has been carried out monthly in the North Atlantic Ocean and North Sea since 1948.

The Global Ocean Ecosystem Dynamics program sponsored by the National Science Foundation and the National Oceanic and Atmospheric Administration is designed to study the effects of physical oceanic conditions and climate on zooplankton, fish recruitment, and adult fish populations. The international Global Ocean Observing System will monitor many relevant features of the ocean and provide data to help improve global stock assessments. These data include standard hydrographic data (e.g., sea-surface temperature, current velocities) and biological information about the distribution and abundance of larval forms. The data can be incorporated into circulation models, and indices of advective losses from the population can be obtained. The indices can be combined with standard stock assessment techniques plus estimates of larval mortality to determine the expected range of fish abundance. This could help fishery managers make better decisions by having a better idea of likely year-class abundance. These approaches are now used operationally as one tool to manage walleye pollock in the western Gulf of Alaska (Megrey et al. 1996, Herrmann et al. 1996). At present, the indices of advection are only qualitative (e.g., large, medium, small). Longer data series, coupled with transport models that have been more thoroughly tested, hold promise of allowing quantitative advection indices. In addition, pre-recruit surveys often are useful. For example, they are good predictors of recruitment of haddock into the fishery on Georges Bank, and are routinely used (S.A. Murawski, National Marine Fisheries Service, Woods Hole, Mass., personal communication, 1998).

Other types of monitoring that contribute to an understanding of marine ecosystems include satellite and in situ observations of oceanic conditions that enable estimations and predictions of physical, chemical, and biological factors that influence fisheries. The system used to monitor the El Niño/Southern Oscillation (ENSO) is a good example. Another potentially useful source of information is the long time-series stations maintained by the Joint Global Ocean Flux Study off the coasts of Bermuda and Hawaii. Monitoring and assessment of fish populations are a potential area of cooperation between fishery scientists and fishers.

New techniques for ocean observations will be available in the near future. Synoptic information on sea-surface characteristics, currents, and bottom habitat and topography are available from a variety of remote-sensing techniques. Optical and acoustical imaging techniques will be used more routinely in the future to make biological measurements and to study characteristics of water masses.

Knowledge of the genetic characteristics of fish populations will allow for better studies of population structure and diversity, gene-flow rates among populations, and the migration of species throughout their ranges. Likewise, fish tags are now available to study fish migrations and mixing of populations, the influence of environmental factors on the movement of individual fish, habitat condition in marine ecosystems, and chemical contamination. These archival tags can measure and store information about the environment (e.g., temperature, depth, irradiance) and fish behavior for up to four years. Analysis of hard parts for various isotopes in bony material laid down during growth is also used to study the migrations of Atlantic bluefin tuna and identify different stocks (Calaprice 1986). Investment in acquiring better information will not only improve assessments of fish stocks but also enhance our ability to characterize and manage marine ecosystems.

Other Scientific Tools

A variety of general tools will facilitate the scientific goals related to managing fisheries in an ecosystem context. These include laboratory and field methods ranging from molecular techniques to whole-system experiments to help understand interactions, the importance of uncertainty, and the interplay of multiple stresses on ecosystems. Improved conceptual understanding of ecosystem organization and functioning and of the relationship between biodiversity and the dynamics of component populations will also contribute to the overall success of ecosystem management.

Much of the present scientific effort related to fisheries responds to short-term tactical management needs, because of a lack of resources to proceed beyond the science needed to fulfill legal and regulatory requirements. Although stock assessments are important, emphasis on them leaves few resources for long-term activities that are necessary for constructing more realistic models of fisheries in an ecosystem context. Different kinds of information and/or more information will be needed to develop longer-term strategic fishery management plans and, in many cases, rehabilitation strategies. It is especially important to understand regime shifts and alternative stable states of marine ecosystems and their component fish species.

Modeling

Obtaining reliable data from observations and experimentation is necessary to develop realistic ecosystem models. Such models, in turn, can indicate new observations that should be collected, new time and space scales for such observations, and new research approaches. At present, a number of different conceptual and mathematical models are used to guide fishery management (see NRC 1998a for a review of stock-assessment models).

Models of biota and their interactions in the complex marine environment will always be imperfect because of imperfect knowledge. Every parameter in every model will be uncertain to some degree. This type of uncertainty is straightforward to handle; even analyses as simple as those describing propagation of errors can be used, whereas in more complicated situations more extensive sensitivity analyses may be warranted. The mathematical descriptions used in existing models may be incomplete. For example, the effects of entire trophic levels may be grossly aggregated, or neglected altogether, as is often the case for higher predators in models focused on plankton. Important details on the distribution of ages or life history stages are sometimes omitted. Environmental variability and uncertainty are rarely addressed adequately in the models of any fishery. In particular, information regarding the largest known source of interannual variability, the ENSO phenomenon, should be incorporated into models of fish populations and marine ecosystems. This could be of enormous help in managing marine ecosystems in the Pacific Ocean region; similarly, information on the North Atlantic Oscillation would help management in the North Atlantic.

Despite the prospect of more comprehensive models and better data to use in such models, it is possible that accurate fishery forecasts will never be achieved for more than a few years in advance because of the chaotic nature of marine systems (Acheson 1995). There are also modeling results that indicate that in an ecosystem composed of as few as five fish species, in which small fish of every species are eaten by larger fish, the biomass of individual species can vary unpredictably even though total biomass remains constant (Wilson et al. 1991).

Multispecies Models and Management

Recognizing that single-species management fails to embrace a realistic ecological perspective, scientists and managers have increasingly promoted the concepts of multispecies (e.g., Sissenwine and Daan 1991) or ecosystem (Sherman et al. 1993) models to supplement assessments made using single-species models. Early models of fishery ecosystems and multispecies fisheries (e.g., Andersen and Ursin 1977) demonstrated the complexity of potential interactions among species and resulting management. Subsequent model building, which adopted more limited multispecies objectives, has achieved a degree of success in understanding both species interactions and the probable consequences of management alternatives (Murawski 1991).

Institutional inertia and allocation problems, the need for more information for multispecies than for single-species models, and some cases of good performance of single-species approaches leave most fisheries under traditional single-species management. In multispecies analysis, information about abundance, coincidence of distribution, diet overlap, consumption rates, maturity, growth, catch at age, and other factors is used to analyze various forms of predatory and competitive relationships among species included in the model.

Multispecies virtual-population analysis (MSVPA), an extension of the single-species VPA that is widely used as a stock-assessment tool, recognizes the importance and complexity of species interactions by accounting for predator-prey relationships among included species (Sparre 1991, Magnusson 1995). Application of MSVPA has demonstrated that natural mortality rates of fish, especially when young, are much higher than previously estimated and that these rates vary from year to year in response to changes in the relative abundances of predators and prey. In single-species VPA, increased mesh sizes or other regulations that increase the average size and age of fish caught are generally predicted to increase overall yields. However, MSVPA models may predict decreased yields for some species if large meshes allow predator populations to expand, thus increasing their consumption of young fish (Magnusson 1995). This demonstrates the usefulness of such models for investigating ecosystem processes and for developing management strategies. However, multispecies models require more information for success than single-species models, and should be considered as supplements rather than replacements for them.

Another modeling approach deals with trophic interactions among the living elements of marine ecosystems, and drawing inferences about possible ecosystem responses to exploitation of various components from the structure and behavior of such models. Such modeling provides an opportunity to use ecosystem approaches for fishery management. The trophic modeling approach has been used for marine systems because of its ability to incorporate disparate data on the biomasses and feeding interactions of marine organisms to enable rigorous system descriptions, comparisons, and inferences related to fishery impacts on ecosystems (Sissenwine et al. 1984, Pauly and Christensen 1995, Christensen and Pauly 1998). Further, once such a model has been constructed for an ecosystem, and the balance of trophic flows established among its elements, dynamic simulations can be run. These simulations can be used to explore first-order effects of various management interventions at all levels in the ecosystem (Walters et al. 1997).

Two observations regarding the use and limitations of existing and potential models are appropriate. First, ecologists and fishery scientists have developed a great variety of models that yield a range of predictions of population and ecosystem variables. The variation of predictions from comparable models could provide an estimate of the level of uncertainty in our understanding of pertinent marine ecosystems. Second, some surprises might never be captured in models. Such surprises might include economic downturns, cultural shifts, or socio-political upheavals. For example, the dissolution of the Soviet Union has had a major impact on fisheries for Antarctic krill and other species because the combined fishing activities of the Confederation of Independent States are less than the previous effort of the Soviet Union. Perhaps economic simulations could account for small perturbations of the status quo, but predicting the most important and largest disturbances is unlikely.

Experimentation and Adaptive Management

Fisheries are large-scale perturbations that provide the opportunity for experimentation at time and space scales that would never be supported by any of the usual funding agencies. The need for experimentation and some of the difficulties inherent in experiments with marine fisheries have been described elsewhere (e.g., Larkin 1972; Pikitch 1988; McAllister and Peterman 1992; Policansky 1993a, 1993b). Indeed, the oft-repeated advice to use adaptive management (e.g., Walters 1986; NRC 1996a, 1996b) is advice to take an experimental approach, one in which the management regimes are *designed* to facilitate data collection and—even more important—hypothesis testing. Thus, the management plan evolves as hypotheses are either supported or rejected. Indeed, the discussion of rights-based allocation schemes above makes clear how important experimentation is, both in the natural and the social sciences.

In many cases, fishing has been going on for so long that experiments are difficult because the results have already occurred (e.g., Policansky 1993a, 1993b; Auster et al. 1996). But sometimes a long history of fishing can be an advantage when an unplanned perturbation occurs. For example, when World War II forced a cessation of fishing in the North Sea, the long data set allowed a careful analysis of its effects (Beverton and Holt 1957, Rijnsdorp 1992). Not surprisingly, there was a large increase in many fish populations.

Marine protected areas also can provide a great deal of information on the effects of fisheries, environmental fluctuations, and other factors on fishing if they are implemented adaptively (see, e.g., the study of Polovina and Haight [in press] described above). That requires that carefully designed monitoring programs accompany the implementation of protected areas. Similarly, the introduction of new fishery regulations, such as bycatch-reduction devices or turtle-excluder devices, provides opportunities for developing and testing ecological ideas as well as learning about the effects of fishing and the effectiveness of fishery management.

Deliberate experimentation with a public resource or profits is not lightly undertaken (Policansky 1993b) because there are risks involved. One is that adaptive management can take a very long time to mature into a successful program (Walters et al. 1993). Another is that populations or ecosystems could be adversely affected by experimental overfishing. But the committee presumes that the need for useful information in general outweighs the risks, although the latter must be carefully considered. For the Bering Sea, the NRC actually recommended deliberate overfishing in restricted areas as a way of gaining information about the ecosystem (NRC 1996a), and this committee agrees that useful information could be gained in that way if experimental fishing is carefully controlled.

ECOSYSTEM-BASED APPROACHES TO MANAGING FISHERIES

Fish populations are one portion of complex ecosystems that are affected by many natural and human-induced factors. Marine ecosystems are used by individuals with a range of perspectives and attitudes about them and about fisheries specifically. Other people rely on fish only as a food product and never visit the ocean. Some individuals depend on the ocean for their livelihood, via fisheries or some unrelated commercial activities. Still others value marine ecosystems and fish populations mostly from an aesthetic perspective. This variety of perspectives invariably produces conflicting goals regarding uses of marine ecosystems. It is one reason there have been so many calls in recent years for ecosystem-based approaches to resource management.

It is the perception of many observers that single-species fishery management has failed (Ludwig et al. 1993, Safina 1995) and that a new approach, which recognizes ecosystem values, is required to achieve sustainable fisheries. A move toward fishing and management that recognize the importance of species interactions, conserve biodiversity, and permit utilization only when the ecosystem or its productive potential is not damaged is a worthy objective. This chapter concludes with a brief discussion of just how such an approach can be applied to fishery management.

It is clear that not enough is known about most large marine ecosystems to implement a reliable whole-ecosystem approach to management. In any case, it is probably beyond our capabilities to manage some aspects of marine ecosystems, such as ENSO events and large-scale migrations. Not enough is known about trophic relationships, environmental variability, community interactions, migration patterns, and many other factors to "manage the ecosystem" as one might try to manage a terrestrial game reserve or a farm. What, then, does an ecosystem-based approach provide to fishery management that is lacking from a single-species approach or even from the multispecies approaches that have been discussed for decades?

The best answer at present is that it is uniquely useful in helping to set policy frameworks that include fish production and ecosystem goods and services; it acknowledges the critical role of ecosystem processes and a much broader focus than only the species of concern for fishing. It acknowledges that humans depend on these ecosystems for a suite of services as well as goods. An ecosystem approach includes a recognition that many segments of society have many goals and values with respect to marine ecosystems and that pursuit of any one goal is likely to affect how well other goals can be achieved. It does *not* provide an excuse for ignoring the biology and economics of individual species, industries, interest groups, and other segments of society. The committee concludes that an ecosystem approach, so defined, and adopted in accordance with the recommendations outlined in this report, will improve prospects for the sustainability of marine fisheries. The approach is described in more detail below.

A Suggestion for Action

An ecosystem approach to fishery management addresses human activities and environmental factors that affect an ecosystem, the response of the ecosystem, and the outcomes in terms of benefits and impacts on humans. The essence of the framework is characterized by stresses, responses, and benefits. The traditional view of a fishery narrowly fits into this framework with fishing as the only stress, the ecosystem response specified solely in terms of the effect of fishing mortality on a single species, and the outcome in terms of catch. One way of achieving an ecosystem approach is to incrementally add to the list of stresses, the scope of the ecosystem responses, and the type of benefits considered in fishery management.

Additional stressors might be forms of degradation in habitat and environmental quality. Experiments on the biological response of the resource (the exploited species) to these stresses would allow the stresses to be taken into account in population models that are traditionally used to determine the effect of fishing. In this way, several different forms of population stress can be compared quantitatively to the stress of fishing. The comparison would be helpful because much is known about how populations respond to fishing but less about population responses to other forms of stress. Even if managers decide not to experiment by deliberately stressing fisheries, they can instead take advantage of whole-system "experiments" (e.g., wars, major environmental fluctuations) and other management actions as described in the section on adaptive management above.

The type of responses to stresses can also be expanded incrementally by trophically linking species. Multispecies virtual-population analysis is an example of this approach. Ultimately, exploited species need to be linked to other components of ecosystems so that indirect responses to stress can be addressed.

Finally, there is a need for incrementally increasing the scope of benefits from fisheries. Benefits of recreational and subsistence fisheries need to be determined. Nonmonetary benefits of ecosystem services also need to be considered. Methods that express benefits in a common metric need to be applied and improved so that decisions can be made between alternative forms of management of fisheries and other human activities.

This incremental approach will take a long time to evolve to a point where it takes account of all the important stresses, responses, and benefits, but it allows immediate progress by taking advantage of the existing framework that is used in fisheries.

Elements of the Approach

Ecosystem Monitoring. An ecosystem approach to fishery management requires a long-term commitment to systematic and carefully standardized observations of the state of ecosystem components,[3] including measurements of stresses and benefits. Traditionally, only commercially exploited species and catches have been monitored. There are notable exceptions such as the California Cooperative Oceanographic Fisheries Investigation (CalCOFI) off California, Continuous Plankton Recorder surveys in the North Atlantic, and the National Oceanic and Atmospheric Administration's Marine Resources Monitoring, Assessment and Prediction program off the Northeastern United States. These programs monitor components of the plankton community and some environmental variables. An ecosystem approach will require expansion of these programs. As an example, consideration should be given to using bycatch data as a source of information about the ecosystem where practicable. Some degree of calibration could be provided by traditional sampling approaches.

Monitoring Human Systems. Much information is also needed on the behavior of people and their social, economic, and political institutions. As described above, information is needed on community-based management; matching institutional scales of complexity to biological ones; the responses of individuals and institutions to a variety of economic, environmental, and political factors; how and whether ITQ and related management regimes can be designed to realize the benefits of market-based approaches to exploitation and stewardship; and so on. Humans are part of the ecosystem and an ecosystem-based approach to management requires information on humans and their systems as well as on the other parts of marine ecosystems.

Application of Ecosystem Principles. While knowledge of how ecosystems function is still very incomplete, much is known about the structural characteristics of ecosystems, how structure relates to functioning, and how structure and functioning respond to various types of stress. For example, the role that top predators play in stabilizing ecosystems has been studied extensively. However incomplete our knowledge about ecosystem principles, what is known should be given formal consideration in fishery-management decisions.

Traditionally, fisheries have been managed by controls on the catch or the amount of fishing activity. There have also been controls on the time and place where fishing is allowed to protect certain components of the fishery resource (spawners) or to make it harder to catch fish so as to protect the resource from the fishery. These approaches have rarely addressed the broader ecological implications of fishing, and for this reason, new methods are needed. One new approach that is receiving increasing attention is marine protected areas, as discussed earlier.

Cross-Sectoral Institutional Arrangements. Marine ecosystems are affected by many human activities in addition to fishing. Traditionally, different institu-

---

[3]The boundaries of marine ecosystems are hard to define, and many species are so migratory that they use more than one ecosystem during their lives, however an ecosystem is defined. Thus an ecosystem-based approach will sometimes require that attention be paid to more than one ecosystem at the same time.

tions, such as government agencies, have had responsibility for managing these activities. Institutional arrangements are needed that require decision makers to consider the effects of one sector's actions on other sectors, such as the effects of agriculture on water quality.

Large Marine Ecosystem Approach (LME). In the past decade there have been a series of meetings classifying coastal areas of the world's ocean into LMEs, and promoting an ecosystem approach to studying and managing these ecosystems. Recently, international donor agencies have shown interest in funding regional programs based on an LME approach. The approach identifies five modules that need to be addressed, including (1) productivity (i.e., the base of the food chain), (2) fishery resources, (3) ocean health (amount and quality of habitat), (4) socioeconomics, and (5) governance. Monitoring strategies for the first two modules are most developed, but the approach stresses the importance of all five modules in order to properly manage ecosystems. The LME approach incorporates several of the elements of an ecosystem approach to fishery management discussed in this report (Sherman et al. 1990, 1993).

The Precautionary Approach. Recently, this approach has gained acceptance for the management of fisheries, as indicated in the United Nations Agreement for Straddling Stocks and Highly Migratory Species. Perhaps it is even more applicable to an ecosystem approach to fishery management than it is to traditional single-species management, because of the level of uncertainty about ecosystems and the potential risks associated with their misuse. It seems unlikely that sustainability of marine fisheries will be achieved without a more pervasive and stronger commitment to the precautionary approach.

6

Conclusions and Recommendations

CONCLUSIONS

Many populations and some species of marine organisms have been severely overfished. Fished and unfished populations have been affected by other human activities, such as coastal development, as well. Those populations and species are ecosystem components and consume or provide significant fractions of the ecosystem's production. Fishing thus affects not only exploited species but also other species that are linked ecologically or environmentally with fished species and their ecosystems. In addition, many current fishery problems are the legacy of a misplaced belief in the inexhaustibility of marine resources, which led to management that did not create incentives for conservation. As a result, many species have been overexploited and more are at risk; there is severe overcapacity of fishing power, which puts pressure on managers to make risk-prone decisions, and as a result many marine fisheries under current management practices are not sustainable at societally acceptable levels.

The committee concludes that a significant overall reduction in fishing mortality is the most comprehensive and immediate ecosystem-based approach to rebuilding and sustaining fisheries and marine ecosystems. The committee's specific recommendations, if implemented, would contribute to an overall reduction in fishing mortality, which is required to rebuild populations, reduce bycatch and discards, and reduce known and as-yet-unknown ecosystem effects.

Earlier chapters in this report describe many difficulties that have contributed to the current overexploited state of the world's marine fisheries. They include risk-prone management, political disagreements and lack of commitment to resource conservation, inappropriate socioeconomic rewards resulting from

117

ill-defined property rights, overcapitalization and excess fishing capacity, inadequate statistics and scientific information, lack of attention to whole ecosystems or to nonfished ecosystem components, lack of predictability owing to environmental and other fluctuations, and mismatches between the time and space scales of fisheries (including fishers) and management institutions. The factors are not mutually independent; several of them derive from the existence of others. It is impossible with present knowledge to assign relative weights to the contributions of those factors to the overall problem and perhaps it always will be. However, the factors do provide a framework for recommendations to improve the sustainability of marine fisheries. Therefore, the committee has focused on recommendations that are likely to improve the sustainability of marine fisheries, whatever the causes of the current difficulties. It has also tried to emphasize recommendations that lead to identifiable actions.

RECOMMENDATIONS

The committee recommends the adoption of an ecosystem-based approach for fishery management to reduce overall fishing mortality. Its goal should be to rebuild and sustain populations, species, biological communities, and marine ecosystems at high levels of economic and biological productivity and biological diversity, so as not to jeopardize a wide range of goods and services from marine ecosystems, while providing food, revenue, and recreation for humans. An ecosystem-based approach that addresses overall fishing mortality will reinforce other approaches to substantially reduce overall fishing intensity. It will help produce the will to manage conservatively, which is required to rebuild depleted populations, reduce bycatch and discards, and reduce known and as-yet-unknown ecosystem effects. Although this approach will cause some economic and social pain at first, it need not result in reduced yields in the long term because rebuilding depleted fish populations should offset a reduction in fishing intensity and increase the future sustainable yields.

Adopting a successful ecosystem-based approach to managing fisheries is not easy, especially at a global or even continental scale. That is why the committee's recommendations include incremental changes in various aspects of fishery management. The elements of this approach, many of which have been applied in single-species management, are outlined below. They include assignment of fishing rights or privileges to provide conservation incentives and reduce overcapacity, adoption of risk-averse precautionary approaches in the face of uncertainty, establishment of marine protected areas, and research. These significant steps that must be taken to make ecosystem approaches to fishery management successful are not entirely new concepts, although they are not easy to implement. As such measures are being implemented, more innovative management tools and techniques, such as marine protected areas, "virtual communities," and ecosystem modeling for fishery management, can be tested and implemented. The following recommendations are specific ways to achieve the broad goals outlined above.

Conservative Single-Species Management

Managing single-species fisheries with an explicitly conservative approach would be a large step toward achieving sustainable marine fisheries. A moderate level of exploitation might be a better goal for fisheries than full exploitation, because full exploitation tends to lead to overexploitation. Many species are overfished, even without considering the ecosystem effects of fishing for them. Therefore, the committee recommends that management agencies and decision makers adopt regulations and policies that strongly favor conservative management and penalize overfishing. Recent amendments to the Magnuson-Stevens Fishery Conservation and Management Act call for such an approach. This, of course, is only a step; by itself it does not appear to be enough to achieve sustainable marine fisheries or to protect marine ecosystems.

In implementing this recommendation, managers should be aware of various factors. For example, long-lived, slow-growing species will recover from overfishing much more slowly than short-lived species. Species whose nursery or feeding habitats have been altered or degraded will recover more slowly, if at all, than those with intact habitats. Breeding aggregations are particularly vulnerable to overfishing. Natural fluctuations will influence population sizes. Finally, conservative approaches can have significant socioeconomic effects, and will require political and managerial commitment and support to be effective.

Incorporating Ecosystem Goals Into Management

Explicit management goals should be established for fisheries that take account of the full value of the goods and services of ecosystems. The aim is to sustain the capacity of ecosystems to produce goods and services at all scales (from local to global) and to provide equitable consideration to the rights and needs of all beneficiaries and users of ecosystem goods and services.

To achieve this difficult goal, it is necessary to predict how the levels of goods and services provided by an ecosystem might change when a variety of ecosystem characteristics change naturally or are altered by human action. Such predictions need more field information and better models. Ecosystems in estuaries, continental shelves, the open ocean, coral reefs, and other areas require different types of data and model parameters. For example, both overfishing and pollution need to be considered in urban estuaries. On coral reefs, good spatial information may be available, but species-specific life history data are relatively difficult to obtain. Numerical models should include key biological and physical indices and spatially and temporally explicit relationships and should help in the development of comparable performance measures (e.g., biological, economic, social) to compare ecosystems in terms of the effects of multiple stresses on their capability to produce goods and services function and the benefits to society from multiple ecosystem uses.

Understanding the relationships of fish populations to ecosystem functioning needs to be based on principles and concepts of community ecology. Trophic interactions (see Chapter 3, especially Box 3-1) and networks of competitive interactions need to be understood at more than one spatial and temporal scale. Larval dispersal and the timing of events related to dispersal and settlement need better understanding for most species. The spatial and temporal components of specific ecological relationships relevant to large-scale changes in ecosystem functioning need to be better understood in most regions that support major fisheries. Models should allow the development and application of new indicators of ecosystem functioning and the dynamics of fish populations to permit assessment of management performance. The indicators should relate to community structure, biodiversity, and health, growth, and reproductive potential of individuals in the ecosystem. Goals or targets should be based on the indicator values, and decisions should be keyed to indicator values having a priori action levels. Such indicators should be compared and evaluated with respect to fishery and ecosystem goals and behavior.

There are more ideas than experience in using ecosystem approaches to fishery management. Marine protected areas and adaptive management have demonstrated their effectiveness in some situations (and thus could support implementation now), but new research is needed to develop and extend the use of these tools.

Dealing with Uncertainty

Fisheries are managed in the context of an incomplete understanding of fish-population dynamics, interactions among species, effects of environmental factors on fish populations, and effects of human actions. Therefore, successful fishery management will have to successfully incorporate and deal with uncertainties and errors. Many of the problems facing fishery managers are questions concerning long-term versus short-term goals and benefits, and uncertainty often leads to an emphasis on short-term actions at the expense of long-term solutions. Uncertainties can induce individuals to use a short-term horizon for decisions related to exploitation and investment, and incentive and management structures must counteract these responses to uncertainty that jeopardize sustainability.

Fishery management can be made to incorporate the variability and uncertainty of the real world by changing management goals to account for uncertainty, giving more emphasis to long-term strategic concerns and less emphasis to short-term variations, and developing management tools that are robust to uncertainty. Developing such management tools requires management to recognize uncertainty and variation as an unavoidable part of natural resource management;

instead of trying to reduce the variations, it should try to reduce their adverse effects. Ways to reduce those adverse effects have been described in this report; they include developing more robust institutional structures and procedures, including economic incentives; alternative strategies to control the amount, timing, and spatial distribution of fishing effort; and the use of marine protected areas.

Explicit incorporation of uncertainty into management decisions is increasing. New laws, conventions, and beliefs—for example, the Food and Agriculture Organization's Code of Conduct and the United Nations Conference on Straddling Fish Stocks and Highly Migratory Fish Stocks—require adoption of a precautionary approach. Institutions cannot eliminate uncertainties, but they can reduce the likelihood that these uncertainties will have a serious impact on a fishery and its host ecosystem, especially if they emphasize precautionary approaches and do not rely on the precision of estimates that vary over space and time. Marine protected areas and effort-based and other controls based on relatively invariant aspects of a fishery are less susceptible to measurement and other errors and can be implemented as permanent aspects of a management regime.

Reducing Excess Fishing Capacity

Excess fishing capacity and overcapitalization reduce the economic efficiency of the fisheries and usually are associated with overfishing. Substantial global reductions in fleet capacity are the highest priority for dealing with uncertainty and unexpected events in fisheries and to help to reduce overfishing. However, overcapacity is a symptom of socioeconomic incentive systems and management regimes, not a fundamental property of fisheries. Overcapacity has been created unintentionally by many national and international institutions through lack of property rights, subsidies, and other activities that circumvent market forces.

Fishers adapt ingeniously to regulations designed to reduce fishing capacity, by improving technology, fishing "smarter" or harder, and modifying their techniques. So fishing capacity is difficult to manage directly without also changing other socioeconomic and management incentives. For this reason the committee recommends that managers' primary focus *not* be on direct management of fishing capacity alone. Instead, managers and policy makers should focus on developing or encouraging socioeconomic and other management incentives that reward conservative use of marine resources and their ecosystems and should learn to understand and address the problems of subsidies (see "Socioeconomic Incentives" below). Direct management of fishing capacity is more appropriate in extreme or urgent circumstances or as a first step in establishing a more sustainable system of using marine resources. Then the degree of overcapacity can be used as one indicator of the sustainability of a fishery.

All direct methods of reducing overcapacity will have social costs that need to be evaluated and considered when determining an approach to be used for a specific fishery. To reduce and monitor fishing capacity there is a need for better information about capacity, including fleet size, type of ships and gear, ownership, and status of operation. Reductions in the capacity of a specific fishing fleet should not be allowed to result in capacity increases in other fisheries, either national or international. Whether downsizing should favor small vessels over large ones, or one gear type over another, should be evaluated on a fishery-by-fishery basis. Simple buy-back programs have often been ineffective and even counterproductive in the past when large amounts of money have been spent to buy out the least efficient vessels. If there are no incentives to reduce fishing power further, the remaining individuals may invest additional capital and increase overall fleet capacity.

Marine Protected Areas

Marine protected areas—where fishing is prohibited—have been effective in protecting and rebuilding populations of many (but not all) marine species. They often increase the numbers of fish and other species in nearby waters. Fishery-management agencies in the United States have often approached this option by closing areas to fishing for considerable periods. These and other experiences in the United States and elsewhere lead the committee to recommend the establishment of permanent marine protected areas in appropriate locations adjacent to all U.S. coasts.

It is important that productive areas—that is, areas in which fishing is good or once was—be protected for this management approach to have the greatest effectiveness. This is because the productive areas have greater potential for rebuilding than less-productive areas. To be effective, protected areas should be established for species whose behavior depends to some degree on structure— that is, species that live, breed, feed, or take shelter on or around the topography of the coast or the bottom of the ocean. They will be most effective for species whose entire life cycle is spent in association with structure or whose juveniles are largely confined to the protected area. Wholly or largely pelagic species move according to ocean currents and thus are likely to benefit less than other species from fixed protected areas.

The design and implementation of marine protected areas should involve fishers so that they believe the resulting systems will protect their long-term interests as well. Involvement of fishers will also provide operational integrity. Attempts to develop marine protected areas in the United States have been strongly opposed by some fishers, so this is a key strategy.

Marine protected areas that allow certain types of catches or other uses (e.g., multiple-use management zones) may serve as an initial step in creating more exclusive reserves. Multi-use zones are often used as a way to allocate an available ocean area to allow for varying levels of use and to maximize synergies among uses while keeping those activities that may interfere with one another separate.

Protected areas of almost any size have some potential to be useful, but to have significant effects the total area protected must be a substantial portion of the potential fishing area. The committee cannot specify what percentage should be protected before the results of many current, proposed, and planned research activities are available, but, based on current theory and experience (as described in Chapter 5), a much greater portion of potential fishing area needs protection. Recent calls for protecting 20 percent of the potential fishing area provide a worthwhile reference point for future consideration and emphasize the importance of greatly increasing the area protected. Increasing the area of marine environments receiving such protection should be considered in the context of enforcement requirements, other management approaches, and the loss of revenues and ecosystem services likely to result from a continuation of current practice. Marine protected areas are not alternatives to other methods of fishery management—they will not work that way—but instead are one major tool among many important ones for protecting ecosystems and achieving sustainable fisheries. For marine protected areas to be most effective as fishery-management tools, their intended purposes must be clearly defined.

Bycatch and Discards

Bycatch and discards should be considered as part of fishing activities rather than only as side effects of them. This means that estimates of bycatch should be incorporated into fishery-management plans and taken into account in setting fishing quotas and in understanding and managing fishing to protect ecosystems and nonfished ecosystem components. In some cases, allocating individual transferable quotas for bycatch shows promise, rather than only setting fleetwide or fishery-wide quotas. The committee recommends the adoption of individual bycatch quotas where appropriate, perhaps on an experimental basis in fisheries where information is lacking. This approach has the advantage of specifying a result and allowing industry the flexibility to choose the method of achieving that result. In some cases, technological developments and careful selection of fishing gear (e.g., bycatch-reduction devices) have been effective in reducing bycatch, and those options should be considered and developed where appropriate. Reduction of effort in some areas or at some times might be needed to reduce bycatch.

Much more information is needed on discards and on bycatch and its fate (i.e., whether bycatch is retained or discarded). Bycatch and discards in recreational fisheries can be significant in some places, and much more information is needed on recreational bycatch and discards as well. In implementing this recommendation, managers need to pay attention to the possibility that bycatch reduction might displace some fishing effort to other fisheries or other areas in undesirable ways.

Technology

Fishing and processing technology has been evolving ever faster since the nineteenth century, and there is no reason to expect that evolution to stop or slow down. It is unpredictable, as technological innovation often is, and technological innovations developed or used outside the fishing industry (e.g., railroads, internal combustion engines, onboard refrigeration, electronic navigation systems, electronic communication and trading systems) often influence total fishing effort and its distribution in time and space. Most technology is developed outside management agencies, and so agencies are unlikely to find it easy to keep abreast of it. Therefore, instead of trying to have regulations keep abreast of technology, to the degree possible, managers should encourage management and incentive regimes that favor conservation, whatever technology (within reason) is used. One example of this approach might be to consider regulating catch size and composition in some circumstances rather than gear, and let the fishers develop the appropriate gear. Monitoring would still be required to check the effectiveness of the gear and its effects on other ecosystem components.

Institutions

Too often, fishery-management institutions do not operate at time and space scales that match those of important processes that affect fisheries. It is therefore important to adapt institutional structures—building on their many strengths—so as to improve the match of time and space scales. Political boundaries, particularly state and national boundaries, often complicate management. Successful models, such as the International Pacific Halibut Commission, are worth emulating in other similar cases.

Management structures that include many relevant groups of stakeholders, like the regional fishery-management councils in the United States, are more likely to be successful in pluralistic societies than those that exclude important groups of stakeholders. The challenge is to develop structures that incorporate diverse views without being compromised by endless negotiations or conflicts of interest.

The committee endorses the advice of Miles (1994) to develop institutional structures that

- effectively and equitably reduce excess capacity,
- broaden the focus of fishery management to include all sources of environmental degradation that affect fisheries,
- structure the duty to cooperate and conserve through institutional principles,
- develop and implement effective monitoring and enforcement, and
- have the capacity to mandate collection and exchange of vital data.

To achieve these goals, the spatial and temporal scales at which the institutional structures operate should better match those of important processes that affect fisheries. Participation in management should be extended to all parties with significant interests in the marine ecosystems that contain exploited marine organisms. Effective and equitable management requires clear and explicit goals and objectives.

Socioeconomic Incentives

Because many current socioeconomic incentive systems often encourage or lead through excess capacity to overfishing, it is essential to modify them. The committee concludes that appropriate socioeconomic incentives will be based on clearer definitions and assignments of exclusive (transferable) rights and responsibilities to government, virtual communities, individual entrepreneurs, geographical communities, and other entities. The exclusive rights include individual transferable quotas (ITQs or IFQs), community-development quotas (CDQs), and various approaches to community management. Most of these approaches are fairly new, at least in their implementation, and not enough experience has been gained to make categorical recommendations about them. Also, it is clear that different approaches will be more or less effective in different situations, so an adaptive approach is essential.

The committee concludes that in most cases rights-based approaches are preferable to traditional open-access fishery-management systems, despite the difficulties sometimes associated with them. In particular, the committee recommends experimental approaches to the development of virtual communities (as described in Chapter 5). This would include the experimental establishment of management groups in which participation is based on whether the parties share an interest in the fishery and its associated habitat, with less emphasis than normal given to where they live or their direct financial interest.

Information Needs

This report has described many areas of scientific uncertainty. Those areas include "traditional" fishery science and management, the structure and functioning of marine ecosystems, and social and economic determinants and consequences of fishers' behavior and management programs. Therefore, the committee recommends research in the following areas:

• *Understanding marine ecosystems.* One approach that seems likely to be productive is an effort to understand mechanisms at lower levels of organization (i.e., populations and communities).

• *Long-term data sets obtained through long-term research and monitoring programs are essential bases for adaptive management.* The information needs and prospects described in Chapter 5 reflect the areas that the committee considers to be of the greatest importance.

• *Models.* Promising modeling approaches are described in Chapter 5; they include models that incorporate environmental variability (e.g., ENSO events) into fishery models, multispecies models, and trophic models. These models need further development, testing, analysis, and calibration to varying degrees. Indeed, one of their greatest values is in directing and clarifying research needs. Models can also be used strategically by managers to add an ecosystem perspective to their annual decision making.

• *Socioeconomic information.* Basic social and economic information is needed on all aspects of fishing and the people who engage in it. Much information is needed on the effects and effectiveness of various forms of rights-based management approaches and other management regimes, the way people behave in response to different economic and social incentives, and on barriers to cooperation and sharing of information. The committee particularly recommends research into the concept of virtual communities described in Chapter 5.

Literature Cited

Abbott, I.A., and J.N. Norris (eds.). 1985. Taxonomy of Economic Seaweeds with Reference to Some Pacific and Caribbean Species. California Sea Grant Program. La Jolla, CA.

Acheson, J.M. 1995. Environmental protection, fisheries management, and the theory of chaos. Pp. 155-160 in National Research Council. Improving Interactions Between Coastal Science and Policy: Proceedings of the Gulf of Maine Symposium. National Academy Press, Washington D.C.

Agardy, M.T. 1994. Advances in marine conservation: The role of marine protected areas. Trends in Ecology and Evolution 9:267-270.

Akatsuka, I. 1990. Introduction to Applied Physiology. SPB Academic Publishing bv, The Hague, Netherlands.

Akatsuka, I. 1994. Biology of Economic Algae. SPB Academic Publishing bv, The Hague, Netherlands.

Alaska Sea Grant. 1996. Solving Bycatch: Considerations for Today and Tomorrow. Alaska Sea Grant College Program Report No. 96-03. University of Alaska, Fairbanks.

Albemus, G. 1997. The Patagonian Toothfish and Norwegian Interest. Norwegian Society for the Conservation of Nature, Barents Sea Office, Leines, Norway.

Alcala, A.C. 1988. Effects of marine reserves on coral fish abundances and yields of Philippine coral reefs. Ambio 17:194-199.

Alexander, L.M. 1990. Geographic perspectives on the management of large marine ecosystems. Pp. 220-223 in K. Sherman, L.M. Alexander, and B.D. Gold (eds.). Large Marine Ecosystems: Patterns, Processes and Yields. American Association for the Advancement of Science, Washington, D.C.

Allison, G., J. Lubchenco, and M. Carr. 1998. Marine reserves are necessary but not sufficient for marine conservation. Ecological Applications 8(1) Supplement: S79-S92.

Alverson, D.L., M.H. Freeberg, S.A. Murawski, and J.G. Pope. 1994. A Global Assessment of Fisheries Bycatch and Discards. FAO Fisheries Technical Paper 339, Food and Agriculture Organization, Rome.

Anderson, J.L. 1997. The growth of salmon aquaculture and the emerging new world order of the salmon industry. Pp. 175-184 in E. Pikitch, D.D. Huppert, and M. Sissenwine (eds.). Global Trends in Fisheries Management. American Fisheries Society Symposium 20. American Fisheries Society, Bethesda, Md.

Andersen, K.P., and E. Ursin. 1977. A multispecies extension of the Beverton and Holt theory of fishing, with accounts of phosphorus circulation and primary production. Meddr. Danm. Fish og Havunders 7:319-435.

Annala, J.H. 1996. New Zealand's ITQ system: Have the first eight years been a success or a failure? Reviews in Fish Biology and Fisheries 6:43-62.

Arnason, R. 1996. On the ITQ fisheries management system in Iceland. Reviews in Fish Biology and Fisheries 6:63-90.

Arntz, W.E. 1986. The two faces of El Niño 1982-83. Meeresforschung—Reports on Marine Research 1:1-46.

Arrow, K., B. Bolin, R. Costanza, P. Dasgupta, C. Folke, C. S. Holling, B.-O. Jansson, S.A. Levin, K.-G. Maler, C. Perrings, and D. Pimentel. 1995. Economic growth, carrying capacity, and the environment. Science 268:520-521.

Auster, P.J., R.J. Malatesta, R.W. Langton, L. Watling, P.C. Valentine, C.L.S. Donaldson, E.W. Langton, A.N. Shepard, and I.G. Babb. 1996. The impacts of mobile fishing gear on seafloor habitats in the Gulf of Maine (Northwest Atlantic): Implications for conservation of fish populations. Reviews in Fisheries Science 4(2):185-202.

Ayling, A.M., and A.L. Ayling. 1986. Report to Great Barrier Reef Marine Park. Unpublished. Cited by Roberts and Polunin (1991).

Babcock, J.P., W.A. Found, M. Freeman, and H. O'Malley. 1928. Report of the International Fisheries Commission Appointed Under the North Pacific Halibut Treaty, Number 1. Dominion of Canada, Ottawa. [Unrevised edition published by the International Fisheries Commission, Vancouver and Seattle, 1931.]

Baker, C.S., and S.R. Palumbi. 1994. Which whales are hunted? A molecular genetic approach to monitoring whaling. Science 265:1538-1539.

Baker, C.S., and S.R. Palumbi. 1996. Population structure, molecular systematics and forensic identification of whales and dolphins. Pp. 10-49 in J.C. Avise and J. L. Hamre, eds. Conservation Genetics: Case Histories from Nature. Chapman and Hall, New York.

Baker, C.S., F. Cipriano, and S.R. Palumbi. 1996. Molecular identification of whale and dolphin products from commercial markets in Korea and Japan. Marine Ecology 5:671-685.

Bakun, A. 1993. The California Current, Benguela Current, and Southwestern South Atlantic Shelf ecosystems: A comparative approach to identifying factors regulating biomass yields. Pp. 199-221 in K. Sherman, L.M. Alexander, and B.D. Gold (eds.). Large Marine Ecosystems: Stress, Mitigations, and Sustainability. AAAS Press, Washington, D.C.

Baranov, F.I. 1918. On the question of the biological basis of fisheries. Nauchnyi issledovatelskii ikhtiologicheskii Institut Isvestia 1(1):81-128 (in Russian). English translation by W.E. Ricker, 1945. Mimeographed.

Baumgartner, T.R., A. Soutar, and V. Ferreira-Bartrina. 1992. Reconstruction of the history of Pacific sardine and northern anchovy populations over the past two millennia from sediments of the Santa Barbara Basin, California. CalCOFI Rep. 33:24-40.

Bax, N.J., and T. Laevastu. 1990. Biomass potential of large marine ecosystems: A systems approach. Pp. 188-205 in K. Sherman, L.M. Alexander, and B.D. Gold (eds.). Large Marine Ecosystems: Patterns, Processes and Yields. American Association for the Advancement of Science, Washington, D.C.

Becker, G. 1983. The Fishes of Wisconsin. University of Wisconsin Press, Madison.

Bell, F.H. 1978. The Pacific Halibut: The Resource and the Fishery. Alaska Northwest Publishing Company, Anchorage.

Bell, J.D. 1983. Effects of depth and marine reserve fishing restrictions on the structure of a rocky reef fish assemblage in the north western Mediterranean Sea. Journal of Applied Ecology 20:357-369.

Bengston, J.L. 1984. Review of Antarctic marine fauna. Pp. 1-126 in CCAMLR 1984. Selected Scientific Papers 1982-1984, Part II. Convention on Conservation of Antarctic Marine Living Resources, Hobart, Australia.

Bennett, B.A., and C.G. Attwood. 1991. Evidence of recovery of a surf-zone fish assemblage following the establishment of a marine reserve on the southern coast of South Africa. Marine Ecology Progress Series 75:173-181.

Berkes, F. 1987. Common-property resource management and Cree Indian fisheries in subarctic Canada. Pp. 66-91 in B. McCay and J. Acheson (eds.). The Question of the Commons. University of Arizona Press, Tucson.

Beverton, R., and S. Holt. 1957. On the Dynamics of Exploited Fish Populations. Ministry of Agriculture, Fisheries and Food (UK), Fisheries Investigation (Series 2), Vol 19.

BFAR (Bureau of Fisheries and Aquatic Resources). 1994. 1993 Philippines Fisheries Profile. Fisheries Policy and Economics Division, Bureau of Fisheries and Aquatic Resources, Manila.

Blindheim, J., and H.R. Skjoldal. 1993. Effects of climatic changes on the biomass yield of the Barents Sea, the Norwegian Sea, and West Greenland large marine ecosystems. Pp. 185-198 in K.L. Sherman, M. Alexander, and B.D. Gold (eds.), Large Marine Ecosystems: Stress, Mitigation, and Sustainability. AAAS Press, Washington, D.C.

Bohnsack, J.A. 1982. Effects of piscivorous predator removal on coral reef community structure. Pp. 258-267 in G.M. Caillet and C.A. Simenstad (eds.). 1981 Gutshop: Third Pacific Technical Workshop on Fish Food Habits Studies. Washington Sea Grant, University of Washington Press, Seattle.

Bohnsack, J.A. 1994. How marine fishery reserves can improve reef fisheries. Proceedings of the Gulf and Caribbean Fisheries Institute 43:217-241.

Bohnsack, J.A. 1996. Maintenance and recovery of fishery productivity. Chapter 11, Pp. 283-313 in N.V.C. Polunin and C.M. Roberts (eds.). Tropical Reef Fisheries. Fish and Fisheries Series 20. Chapman and Hall, London.

Bohnsack, J.A. 1998. Application of marine reserves to reef fisheries management. Australian Journal of Ecology 23: 298-304.

Botsford, L.W., J.C. Castilla, and C.H. Peterson. 1997. The management of fisheries and marine ecosystems. Science 277:509-515.

Boynton, W.R., J.H. Garber, R. Summers, and W.M. Kemp. 1995. Inputs, transformations, and transport of nitrogen and phosphorus in Chesapeake Bay and selected tributaries. Estuaries 18:285-314.

Brander, K. 1981. Disappearance of the common skate, *Raia batis*, from Irish Sea. Nature 290:48-49.

Broecker, W.S. 1991. The great ocean conveyor. Oceanography 4(2):79-89.

Broecker, W.S., D.M. Peteet, and R.L. Smith. 1985. Does the ocean-atmosphere system have more than one stable mode of operations? Nature 315:21-26.

Brothers, N. 1991. Albatross mortality and associated bait loss in the Japanese longline fishery in the Southern Ocean. Biological Conservation 55:255-268.

Brubaker, E. 1996. The ecological implications of establishing property rights in Atlantic Fisheries. Pp. 221-251 in B.L. Crowley (ed.). Taking Ownership: Property Rights and Fishery Management on the Atlantic Coast. Atlantic Institute for Market Studies, Halifax, N.S., Canada.

Bublitz, C.G. 1996. Mesh size and shape: Reducing the capture of undersized fish. Pp. 95-99 in Alaska Sea Grant 1996. Solving Bycatch: Considerations for Today and Tomorrow. Alaska Sea Grant College Program Report 96-03, University of Alaska, Fairbanks.

Buck, E. 1995. Overcapitalization in the U.S. Commercial Fishing Industry. Congressional Research Service Report for Congress, 95-296-ENR, Washington D.C.

Buffet, B. 1989. Fishery Overview: Newfoundland Region 1989. Unpublished paper prepared for the Government of Canada Task Force on Northern Cod.

Burkholder, J.M. 1998. Implications of harmful microalgae and heterotrophic dinoflagellates in management of sustainable marine fisheries. Ecological Applications 8(1) Supplement: S37-S62.

Butterworth, D.S., J. A. A. De Oliveira, and K. L. Cochrane. 1993. Current initiatives in refining the management procedure for the South African anchovy resource. Pp. 439-473 in G. Kruse, D.M. Eggers, R.J. Marasco, C. Pautzke, and T.J. Quinn II (eds.). Management Strategies for Exploited Fish Populations. Report AK-SG-93-02, Alaska Sea Grant College Program, Fairbanks.

Buxton, C.D., and M.J. Smale. 1989. Abundance and distribution patterns of three temperate marine fish (Teleostei: Sparidae) in exploited and unexploited waters of the southern Cape coast. Journal of Applied Ecology 26:441-451.

Caddy, J.F., and R.C. Griffiths. 1990. Recent Trends in the Fisheries and Environment in the General Fisheries Council for the Mediterranean (GFCM) Area. Studies and Reviews, General Fisheries Council for the Mediterranean 63. Food and Agriculture Organization of the United Nations, Rome.

Caddy, J.F., and J.A. Gulland. 1983. Historical patterns of fish stocks. Marine Policy 8:267-278.

Calaprice, J.R. 1986. Chemical variability and stock variation in northern Atlantic bluefin tuna. ICCAT Collected Volume of Scientific Papers XXIV (2):222-254 (SCRS/85/36).

California Fish and Game Department. 1997. News release August 4, 1997: "Striped bass summer abundance plummets." California Fish and Game Department, Sacramento.

California Fish and Game Department. 1998. Web site http://delta.dfg.ca.gov/baydelta/monitoring/fishlist.html.

Canada. 1990. Final Report. Northern Cod Review Panel, Ottawa.

Canada. 1995. 1996 Conservation Requirements for Atlantic Groundfish: Report to the Minister of Fisheries and Oceans. Fisheries Resource Conservation Council, Ottawa.

Carlton, J.T. 1989. Man's role in changing the face of the ocean: Biological invasions and implications for conservation of near-shore environments. Conservation Biology 3:265-273.

Carlton, J.T. 1993. Neoextinctions of marine invertebrates. American Zoologist 33:499-509.

Carlton, J.T., and J.B. Geller. 1993. Ecological roulette: The global transport of nonindigenous marine organisms. Science 261:78-82.

Carlton, J.T., J.K. Thompson, L.E. Schemel, and F.H. Nichols. 1990. Remarkable invasion of San Francisco Bay (California, USA) by the Asian clam *Potamocorbula amurensis* I. Introduction and dispersal. Marine Ecology Progress Series 66:81-94.

Carpenter, S.R., and J.F. Kitchell. 1988. Consumer control of lake productivity. Bioscience 38:764-769.

Carpenter, S.R., J.F. Kitchell, and J.R. Hodgson. 1985. Cascading trophic interactions and lake productivity. BioScience 35:634-639.

Carr, M.H., and D.C. Reed. 1993. Conceptual issues relevant to marine harvest refuges: Examples from temperate reef fishes. Canadian Journal of Fisheries and Aquatic Science 50:2019-2028.

Casey, J.M., and R.A. Myers. 1998. Near extinction of a large, widely distributed fish. Science 281:690-692.

Castilla, J.C. 1994. The Chilean small-scale benthic shellfisheries and the institutionalization of new management practices. Ecology International Bulletin 21:47-63.

Castilla, J.C. 1995. The sustainability of natural resources as viewed by an ecologist and exemplified by the fishery of the mollusc *Concholepas concholepas* in Chile. Pp. 153-159 in M. Munasinghe and W. Shearer (eds.). Defining and Measuring Sustainability. The International Bank for Reconstruction and Development, The World Bank, Washington, D.C.

Castilla, J.C., and M. Fernandez. 1998. Small-scale benthic fisheries in Chile: on co-management and sustainable use of benthic invertebrates. Ecological Applications 8(1) Supplement: S124-S132.

Center for Law and Social Policy and The Oceanic Society. 1980. Report of the Southern Ocean Convention Workshop on Management of Antarctic Marine Living Organisms. The Oceanic Society at the Stamford Marine Center, Stamford, Conn.

Chamberlain, G.W. 1997. Sustainability of world shrimp farming. Pp. 195-209 in E. Pikitch, D.D. Huppert, and M. Sissenwine (eds.). Global Trends in Fisheries Management. American Fisheries Society Symposium 20. American Fisheries Society, Bethesda, Md.

Chesapeake Bay Program (CBP). 1995. Chesapeake Bay: Introduction to an Ecosystem. Chesapeake Bay Program, Annapolis, Md.

Chan, F., and R.M. Fujita. 1994. The Pacific Salmon Treaty: Opportunities for Improvement and for Resolving the Current Conflict. Environmental Defense Fund, New York.

Christensen, N., et al. 1996. Report of the Committee on Ecosystem Management of the Ecological Society of America. Ecological Applications 6:665-691.

Christensen, V., and D. Pauly. 1998. Changes in models of aquatic ecosystems approaching carrying capacity. Ecological Applications 8(1) Supplement: S104-S109.

Christy, F.T. 1973. Fishermen quotas: A tentative suggestion for domestic management. Occasional Paper 19, Law of the Sea Institute, University of Rhode Island, Narragansett.

Christy, F.T. 1982. Territorial use rights in marine fisheries: Definitions and conditions. FAO Fisheries Technical Paper No. 227. Food and Agriculture Organization of the United Nations, Rome.

Christy, F.T. 1997. Economic waste in fisheries: Impediments to change and conditions for improvement. Pp. 28-39 in E. Pikitch, D.D. Huppert, and M. Sissenwine (eds.). Global Trends in Fisheries Management. American Fisheries Society Symposium 20. American Fisheries Society, Bethesda, Md.

Clark, J.R., B. Causey, and J.A. Bohnsack. 1989. Benefits from coral protection: Looe Key reef, Florida. Pp. 3076-3086 in O.T. Magoon, H. Converse, D. Miner, L.T. Tobin, and D. Clark (eds.). Coastal Zone 1989: Proceedings of the 6th Symposium on Coastal and Ocean Management. American Society of Civil Engineers, New York.

Clark, C.W. 1980. Restricted access to common-property fishery resources: A game-theoretic analysis. Pp. 117-132 in P. Liu (ed.). Dynamic Optimization and Mathematical Economics. Plenum Press, New York.

Clark, C.W. 1990. Mathematical Bioeconomics: The Optimal Management of Renewable Resources, 2nd ed. Wiley-Interscience, New York.

Clark, C.W. 1996. Marine reserves and the precautionary management of fisheries. Ecological Applications 6(2):369-370.

Clark, C.W., and G.R. Munro. 1982. The economics of fishing and modern capital theory: A simplified approach. Pp. 31-54 in L.J. Mirman and D.F. Spulber (eds.). Essays in the Economics of Natural Resources. North-Holland Press, Amsterdam.

Clark, C.W., and G.R. Munro. 1994. Renewable resources as natural capital. Pp. 343-361 in A.M. Jansson, M. Hammer, C. Folke, and R. Costanza (eds.). Investing in Natural Capital: The Ecological Economics Approach to Sustainability. Island Press, Washington, D.C.

Cloern, J.E., and A.D. Jassby. 1995. Yearly fluctuation of the spring phytoplankton bloom in the South San Francisco Bay—An example of ecological variability at the land-sea interface. Pp. 139-149 in J.H. Steele, T.M. Powell, and S.A. Levin (eds.). Ecological Time Series. Chapman and Hall, London.

Cloern, J.E. 1996. Phytoplankton bloom dynamics in coastal ecosystems: A review with some general lessons from sustained investigation of San Francisco Bay, California. Review of Geophysics 34(2):127-168.

Cohen, A., and J.T. Carlton. 1998. Accelerating invasion rate in a highly invaded estuary. Science 279:555-558.

Cook, R.M., A. Sinclair, and G. Stefánsson. 1997. Potential collapse of North Sea cod stocks. Nature 385:521-522.

Cordell, J. (ed.). 1989. A Sea of Small Boats. Cultural Survival Report 26. Cultural Survival, Inc., Cambridge, Mass.

Costanza, R., R. d'Arge, R. de Groot, S. Farber, M. Grasso, B. Hannon, K. Limburg, S. Naeem, R.V. O'Neill, J. Paruelo, R.G. Raskin, P. Sutton, and M. van den Belt. 1997. The value of the world's ecosystem services and natural capital. Nature 387:253-260.

Crawford, R.J.M. 1991. Factors influencing population trends of some abundant vertebrates in sardine-rich coastal ecosystems. South African Journal of Marine Science 10:365-381.

Crawford, R.J.M., L.J. Shannon, and G. Nelson. 1995. Environmental change, regimes and middle-sized pelagic fish in the southeast Atlantic Ocean. Scientia Marina 59(3-4):417-426.

Cronon, W., Jr. 1983. Changes in the Land: Indians, Colonists, and the Ecology of New England. Hill and Wang, New York.

Crothers, S. 1988. Individual transferable quotas: The New Zealand experience. Fisheries. 13(1):10-12.

Csirke, J. 1988. Small shoaling pelagic fish stocks. Pp. 271-302 in J.A. Gulland (ed.). Fish Population Dynamics, 2nd. ed. John Wiley & Sons, Chichester, England.

Csirke, J. 1995. Fluctuations in abundance of small and mid-size pelagics. Scientia Marina 59:481-490.

Csavas, I. 1994. Important factors in the success of shrimp farming. World Aquaculture. 25(1):35-36.

Cushing, D.H. 1972. Marine Ecology and Fisheries. Cambridge University Press, Cambridge, England.

Cushing, D.H. 1980. The decline of the herring stocks and the gadoid outburst. J. Cons. Int. Explor. Mer 39:70-81.

Cushing, D.H. 1982. Climate and Fisheries. Academic Press, London.

Daily, G. C. (ed.). 1997. Nature's Services: Societal Dependence on Natural Ecosystems. Island Press, Washington, D.C.

Davis, S.K. 1996. Multispecies management: An alternative solution to the bycatch problem. Pp. 251-259 in Alaska Sea Grant 1996. Solving Bycatch: Considerations for Today and Tomorrow. Alaska Sea Grant College Program Report 96-03, University of Alaska, Fairbanks.

Dayton, P.K., S.F. Thrush, M.T. Agardy, and R.J. Hofman. 1995. Viewpoint: Environmental effects of marine fishing. Aquatic Conservation: Marine and Freshwater Ecosystems 5:205-232.

Dugan, J.E., and G.E. Davis. 1993. Application of marine refugia to coastal fisheries management. Canadian Journal of Fisheries and Aquatic Sciences 50:2029-2041.

Duggins, D.O. 1980. Kelp beds and sea otters: An experimental approach. Ecology 61:447-453.

Duggins, D.O., C.A. Simenstad, and J.A. Estes. 1989. Magnification of secondary production by kelp detritus in coastal marine ecosystems. Science 245:170-173.

Duran, L.R., and J.C. Castilla. 1989. Variation and persistence of the middle rocky intertidal community of central Chile, with and without harvesting. Marine Biology 103:555-562.

Dupont, D. 1996. Limited entry fishing programs: Theory and Canadian practice. Pp. 107-128 in D.V. Gordon and G.R. Munro (eds.). Fisheries and Uncertainty: A Precautionary Approach to Resource Management. University of Calgary Press, Calgary, Alberta.

Dye, A.H., G.M. Branch, J.C. Castilla, and B.A. Bennett. 1994. Biological options for the management of the exploitation of intertidal and subtidal resources. Pp. 131-154 in W.R. Siegfried (ed.). Rocky Shores: Exploitation in Chile and South Africa. Springer-Verlag, Heidelberg.

Ebener, M.P. 1997. Recovery of the lake whitefish populations in the Great Lakes. Fisheries 22(7):18-22.

Edwards, S., and S. Murawski. 1993. Potential economic benefits from efficient harvest of New England groundfish. North American Journal of Fisheries Management 13:437-449.

Elmgren, R. 1989. Man's impact on the ecosystem of the Baltic Sea: Energy flows today and at the turn of the century. Ambio 18:326-332.

EPA (U.S. Environmental Protection Agency). 1995a. The First Biennial Progress Report of the Agreement of Federal Agencies on Ecosystem Management in the Chesapeake Bay. Chesapeake Bay Program, U.S. Environmental Protection Agency, Annapolis, Md.

EPA (U.S. Environmental Protection Agency). 1995b. Introduction to an Ecosystem. Chesapeake Bay Program, U.S. Environmental Protection Agency, Annapolis, Md.

Estes, J.A., and J.F. Palmisano. 1974. Sea otters: Their role in structuring nearshore communities. Science 185:1058-1060.

Everett, J.T., A. Krovnin, D. Lluch-Belda, E. Okemwa, H.A. Regier, and J.-P Troadec. 1996. Fisheries. Pp. 511-537 in R.T. Watson, M.C. Zinyowera, and R.H. Moss (eds.). Climate Change 1995: Impacts, Adaptations, and Mitigation of Climate Change: Scientific-Technical Analyses. Contribution of Working Group II to the Second Assessment Report of the Intergovernmental Panel on Climate Change. Cambridge University Press, New York.

Finlayson, A.C. 1994. Fishing for Truth: A Sociological Analysis of Northern Cod Stock Assessments from 1977-1990. Institute of Social and Economic Research, St. John's, Newfoundland.

Fitzpatrick, J. 1995. Technology and fisheries legislation. Unpublished paper presented at the International Technical Consultation on the Precautionary Approach to Capture Fisheries. Lysekel, Sweden, 6-13 June 1995.

Fogarty, M.J., and S. A. Murawski. 1998. Large-scale disturbance and the structure of marine ecosystems: Fishery impacts on Georges Bank. Ecological Applications 8(1) Supplement: S6-S22.

Fogarty, M.J., M.P. Sissenwine, and E.B. Cohen. 1991. Recruitment variability and the dynamics of exploited marine populations. Trends in Ecology and Evolution 6(8):241-246.

Folke, C., and N. Kautsky. 1989. The role of ecosystems for a sustainable development of aquaculture. Ambio 18:234-243.

Folke, C., and N. Kautsky. 1992. Aquaculture with its environment: Prospects for sustainability. Ocean and Coastal Management 17:5-24.

Folke, C., N. Kautsky, H. Berg, Å. Jansson, J. Larsson, and M. Troell. 1998. The ecological footprint concept for sustainable seafood production: A review. Ecological Applications 8(1) Supplement: S63-S71.

Food and Agriculture Organization (FAO). 1992. The State of Food and Agriculture. Food and Agriculture Organization of the United Nations, Rome.

Food and Agriculture Organization (FAO). 1993a. Fishery Statistics: Catches and Landings. 1991. FAO Fisheries Series No. 40, Fisheries Statistics Series No. 111, Vol. 72.

Food and Agriculture Organization (FAO). 1993b. Marine fisheries and the Law of the Sea: A decade of change. Special chapter (revised) of The State of Food and Agriculture, 1992. Food and Agriculture Organization of the United Nations, Rome.

Food and Agriculture Organization (FAO). 1994a. Review of the state of world marine fishery resources. FAO Fisheries Technical Paper 335. Food and Agriculture Organization of the United Nations, Rome.

Food and Agriculture Organization (FAO). 1994b. World Review of Highly Migratory Fish Species and Straddling Stocks, FAO Fisheries Technical Paper No. 337. Food and Agriculture Organization of the United Nations, Rome.

Food and Agriculture Organization (FAO). 1995a. Safeguarding future fish supplies: Key policy issues and measures. Unpublished paper presented at the International Conference on the Sustainable Contribution of Fisheries to Food Security. December 4-9, 1995. Kyoto, Japan. KC/FI/95/1.

Food and Agriculture Organization (FAO). 1995b. The State of World Fisheries and Aquaculture. Food and Agriculture Organization of the United Nations, Rome.

Food and Agriculture Organization (FAO). 1995c. Code of Conduct for Responsible Fisheries. Food and Agriculture Organization of the United Nations, Rome.

Food and Agriculture Organization (FAO). 1995d. Precautionary approach to fisheries, FAO Fisheries Technical Paper 350. Food and Agriculture Organization of the United Nations, Rome.

Food and Agriculture Organization (FAO). 1996a. Fisheries and Aquaculture in Europe: Situation and Outlook in 1996. FAO Fisheries Circular # 911. Food and Agriculture Organization of the United Nations, Rome.

Food and Agriculture Organization (FAO). 1996b. Chronicles of Marine Fishery Landings (1950-1994): Trend Analysis and Potential. FAO Fisheries Technical Paper 359. Food and Agriculture Organization of the United Nations, Rome.

Food and Agriculture Organization (FAO). 1997a. The State of World Fisheries and Aquaculture 1996. Food and Agriculture Organization of the United Nations, Rome.

Food and Agriculture Organization (FAO). 1997b. Commodity Market Review 1996-97. Food and Agriculture Organization of the United Nations, Rome.

Food and Agriculture Organization (FAO). 1997c. World Fisheries Production Shows Slight Increase in 1996. Press Release 97/35. Food and Agriculture Organization of the United Nations, Rome.

Food and Agriculture Organization (FAO). 1997d. A Study of the Options for Utilization of Bycatch and Discards from Marine Capture Fisheries. FAO Fisheries Circular # 928. Food and Agriculture Organization of the United Nations, Rome.

Francis, G.R., J.J. Magnuson, H.A. Regier, and D.R. Talhlem (eds.). 1979. Rehabilitating Great Lakes Ecosystems. Technical Report No. 37. Great Lakes Fishery Commission, Ann Arbor, Mi.

Francis, R.C., and S.R. Hare. 1994. Decadal-scale regime shifts in the large marine ecosystems of the Northeast Pacific: A case for historical science. Fisheries Oceanography 3:279-291.

Garcia, S.M., and C. Newton. 1997. Current situation, trends, and prospects in world capture fisheries. Pp. 3-27 in E. Pikitch, D.D. Huppert, and M. Sissenwine (eds.). Global Trends in Fisheries Management. American Fisheries Society Symposium 20. American Fisheries Society, Bethesda, Md.

Gauvin, J.R., J.M. Ward, and E.E. Burgess. 1993. A description and preliminary evaluation of the wreckfish fishery, *Polyprion americanus*, under individual transferable quotas. Pp. 761-789 in G. Kruse, D.M. Eggers, R.J. Marasco, C. Pautzke, and T.J. Quinn II (eds.). Management Strategies for Exploited Fish Populations. Report AK-SG-93-02, Alaska Sea Grant College Program, Fairbanks.

General Assembly of the United Nations. 1995. Agreement for the Implementation of the Provisions of the United Nations Convention of 10 December 1982 Relating to the Conservation and Management of Straddling Fish Stocks and Highly Migratory Fish Stocks. A/Conf.164/33, August 3.

Ginter, J.J.C. 1995. The Alaska community development quota fisheries management plan. Ocean and Coastal Management 28(1):147-163.

Goode, G.B., and J.W. Collins. 1887. The fresh halibut fishery. Pp. 3-89 in G.B. Goode (ed.). The Fisheries and Fishing Industry of the United States. U.S. Government Printing Office, Washington, D.C.

Goodyear, C.P. 1985. Toxic materials, fishing, and environmental variation: simulated effects on striped bass population trends. Trans. Am. Fish. Soc. 114:107-113.

Goodyear, C.P. 1995. Red snapper in U.S. waters of the Gulf of Mexico. MIA-95/96-05. National Marine Fisheries Service, Southeast Fisheries Center, Miami.

Goodyear, C.P., and P. Phares. 1990. Recent trends in the red snapper fishery of the Gulf of Mexico. CRD-87/88-16. National Marine Fisheries Service, Southeast Fisheries Science Center, Miami.

Gordon, H.S. 1954. The economic theory of common property resources: The fishery. Journal of Political Economy 62:124-142.

Graham, M. 1935. Modern theory of exploiting a fishery, and its application to North Sea trawling. Journal du Conseil 13:264-274.

Graham, M. 1943. The Fish Gate. Faber, London.

Groombridge, B. (ed.). 1992. Global Biodiversity: The Status of the Earth's Living Resources. Chapman and Hall, London.

Gulland, J.A. 1965. Estimation of mortality rates. Annex to the Report of the Arctic Fisheries Working Group. ICES C.M. 1965 (3).

Gulland, J.A. 1972. The Fish Resources of the Oceans. Fishing News Books, Ltd., London.

Gulland, J.A. 1983. Fish Stock Assessment: A Manual of Basic Methods. John Wiley & Sons, Chichester, England.

Haedrich, R.L. 1997. Distribution and Population Ecology. Pp. 79-114 in D.J. Randall and A.P. Farrell, eds. 1997. Deep-Sea Fishes. Academic Press, San Diego, Calif.

Hamre, J. 1994. Biodiversity and exploitation of the main fish stocks in the Norwegian-Barents Sea ecosystem. Biodiversity and Conservation 3:473-492.

Hanna, S. 1995. User participation and fishery management performance within the Pacific Fishery Management Council. Ocean and Coastal Management 28(1-3):23-44.

Hanna, S. 1998. Institutions for marine ecosystems: Economic incentives and fishery management. Ecological Applications 8(1) Supplement: S170-174.

Hart, J.L. 1973. Pacific Fishes of Canada. Fisheries Research Board of Canada Bulletin 180. Ottawa, Ontario.

Herrman, A.J., S. Hinckley, B.A. Megrey, and P.J. Stabeno. 1996. Interannual variability of the early life history of walleye pollock near Shelikof Strait as inferred from a spatially explicit, individual-based model. Fisheries Oceanography 5 (Supplement 1):39-57.

Hixon, M.A., and M.H. Carr. 1997. Synergistic predation, density dependence, and population regulation in marine fish. Science 277:946-949.

Hofmann, E.E., and T.M. Powell. 1998. Environmental variability effects on marine fisheries: Four case histories. Ecological Applications 8(1) Supplement: S23-S32.

Holt, E.W.L. 1895. An examination of the present state of the Grimsby trawl fishery with especial reference to the destruction of immature fish. Journal of the Marine Biological Association (UK) 5:337-447.

Houde, E.D., and E.S. Rutherford. 1993. Recent trends in estuarine fisheries: Predictions of fish production and yields. Estuaries 16(2):161-175.

Hughes, T.P. 1994. Catastrophes, phase shifts, and large-scale degradation of a Caribbean coral reef. Science 265:1547-1551.

Hutchings, J., and R.A. Myers. 1994. What can be learned from the collapse of a renewable resource, Atlantic cod, *Gadus morhua*, off Newfoundland and Labrador? Canadian Journal of Fisheries and Aquatic Sciences 51:2126-2146.

Hutchings, J.A., C. Walters, and R.L. Haedrich. 1997. Is scientific inquiry compatible with government information control? Canadian Journal of Fisheries and Aquatic Sciences 54:1198-1210.

Hviding, E., and G.B.K. Baines. 1994. Community-based fisheries management, tradition and the challenges of development in Marovo, Solomon Islands. Development and Change 25:13-19.

Idyll, C.P. 1978. The Sea Against Hunger. Thomas U. Crowell Company, New York.

International Council for Exploration of the Sea (ICES). 1995. Report of the Study Group on Unaccounted Mortality in Fisheries. ICES CM 1995/B:1. ICES, Copenhagen.

Institute of Marine Research, Bergen. 1995. Marine Resources in Norwegian Waters. (English Summary.) Annual Report of the Institute of Marine Research. Institute of Marine Research, Bergen, Norway.

Iverson, R.L. 1990. Control of marine fish production. Limnology and Oceanography 35:1593-1604.

Jackson, J.J. 1997. Reefs since Columbus. Coral Reefs 16 Supplement:S23-S32.

Jentoft, S. 1989. Fisheries co-management: Delegating government responsibility to fishermen's organizations. Marine Policy 13:137-154.

Jentoft, S., and B. McCay. 1995. User participation in fisheries management: Lessons drawn from international experiences. Marine Policy 19(3):227-246.

Johannes, R.E. 1977. Traditional law of the sea in Micronesia. Micronesia 13:121-127.

Johannes, R.E., and M. Riepen. 1995. Environmental, Economics, and Societal Implications of the Live Reef Fish Trade in Asia and the Western Pacific. South Pacific Forum Fisheries Agency and The Nature Conservancy, Arlington, Va.

Johnson, H.M. 1995. 1995 Annual Report on the United States Seafood Industry. H.M. Johnson & Associates, Bellevue, Wash.

Kawasaki, T. 1992. Mechanisms governing fluctuations in pelagic fish populations. South African Journal of Marine Science 12:873-879.

Kennelly, S.J., and M.K. Broadhurst. 1996. Fishermen and scientists solving bycatch problems: Examples from Australia and possibilities for the northeastern United States. Pp. 121-128 in Alaska Sea Grant 1996. Solving Bycatch: Considerations for Today and Tomorrow. Alaska Sea Grant College Program Report 96-03, University of Alaska, Fairbanks.

Kerr, S.R., and R.A. Ryder. 1997. The Laurentian Great Lakes experience: A prognosis for the fisheries of Atlantic Canada. Canadian Journal of Fisheries and Aquatic Sciences 54:1190-1197.

Kessler, W.B., and H. Salwasser. 1995. Natural resource agencies: Transforming from within. Pp. 171-187 in R.L. Knight and S.F. Bates (eds.). A New Century for Natural Resources Management. Island Press, Washington, D.C.

Klima, E.F., G.A. Matthews, and F.J. Patella. 1986. Abundance and distribution of pink shrimp in and around the Tortugas Sanctuary 1981-1983. North American Journal of Fisheries Management 6:301-310.

Knight, R.L. 1997. Successful interagency rehabilitation of Lake Erie walleye. Fisheries 22(7):16-17.

Kock, K-H. 1992. Antarctic Fish and Fisheries. Cambridge University Press, New York.

Kurien, J. 1998. Traditional ecological knowledge and ecosystem sustainability: On giving new meaning to Asian coastal proverbs. Ecological Applications 8(1) Supplement: S2-S5.

Laevastu, T. 1993. Marine Climate, Weather and Fisheries. Halsted Press, John Wiley & Sons, New York.

Lampman, B.H. 1946. The Coming of the Pond Fishes. Binford and Morts Publishing, Portland, Oreg.

Larkin, P.A. 1972. A confidential memorandum on fisheries science. Pp. 189-197 in B. Rothschild, ed. World Fisheries Policy: Multidisciplinary Views. University of Washington Press, Seattle.

Lauck, T., C.W. Clark, M. Mangel, and G.R. Munro. 1998. Implementing the precautionary approach in fisheries management through marine reserves. Ecological Applications 8(1) Supplement:S72-S78.

Laws, R.M. 1985. The ecology of the Southern Ocean. American Scientist 73:26-40.

Leal, D.R. 1996. Community-run fisheries: Preventing the tragedy of the commons. Pp. 182-220 in B.L. Crowley (ed.). Taking Ownership: Property Rights and Fishery Management on the Atlantic Coast. Atlantic Institute for Market Studies, Halifax, Nova Scotia, Canada.

Lee, K.N. 1993. Compass and Gyroscope: Integrating Science and Politics for the Environment. Island Press, Covelo, Calif.

Leet, W.S., C.M. Dewees, and C.W. Haugen (eds.). 1992. California's Living Marine Resources and Their Utilization. California Sea Grant Report UCSGEP-92-12, University of California, Davis.

Leffler, M. 1993. The striped bass success story: A model for fisheries management? Maryland Sea Grant Program. Marine Notes (Nov.):1-3.

Lenihan, H.S., and C.H. Peterson. 1998. How habitat degradation through fishery disturbance enhances impacts of hypoxia on oyster reefs. Ecological Applications 8(1):128-140.

Levin, S.A. 1988. Sea otters and nearshore benthic communities: A theoretical perspective. Pp. 202-209 in G.G. Van Blaricom and J.A. Estes (eds.). The Community Ecology of Sea Otters. Springer-Verlag, Berlin.

Levy, K., R.M. Fujita, and T.F. Young, 1995. Designing restoration of the San Francisco Bay-Delta-River Ecosystem—Framework for developing ecological indicators and thresholds. Discussion paper for the workshop "Restoration of the San Francisco Bay-Delta-River Ecosystem: Choosing Indicators of Ecological Integrity." Environmental Defense Fund, Oakland, Calif.

Lichtkoppler, F.R. 1997. Ohio's Lake Erie charter fishing industry: 1985-1994. Fisheries 22(1):14-21.

Lluch-Belda, D., R.J.M. Crawford, T. Kawasaki, A.D. MacCall, R.H. Parrish, R.A. Schwartziose, and P.E. Smith. 1989. World-wide fluctuations of sardine and anchovy stocks: The regime problem. South African Journal of Marine Science 8:195-205.

Loverich, G.F. 1996. Thinking beyond the traditional codends. Pp. 101-106 in Alaska Sea Grant 1996. Solving Bycatch: Considerations for Today and Tomorrow. Alaska Sea Grant College Program Report 96-03, University of Alaska, Fairbanks.

Lowry, N., G. Sangster, and M. Breen. 1996. Cod-End Selectivity and Fishing Mortality. Danish Institute for Fisheries Technology and Aquaculture and Scottish Agriculture, Environmental, and Fisheries Department. Study Contract No. 1994/005.

Lu, X. 1997. Aquaculture in China and its effect on global markets. Pp. 185-194 in E. Pikitch, D.D. Huppert, and M. Sissenwine (eds.). Global Trends in Fisheries Management. American Fisheries Society Symposium 20. American Fisheries Society, Bethesda, Md.

Lubchenco, J., A.M. Olson, L.B. Brubaker, S.R. Carpenter, M.M. Holland, S.P. Hubbell, S.A Levin, J.A. MacMahon, P.A. Matson, J.M. Melillo, H.A. Mooney, C.H. Peterson, H. R. Pulliam, L.A. Real, P.J. Regal, and P.G. Risser. 1991. The Sustainable Biosphere Initiative: An ecological research agenda. Ecology 72:371-412.

Ludwig, D., R. Hilborn, and C. Walters. 1993. Uncertainty, resource exploitation, and conservation: Lessons from history. Science 260:17 and 36.

Mac, M.J., and M. Gilbertson, eds. 1990. Proceedings of the Roundtable on Contaminant-Caused Reproductive Problems in Salmonids, held Windsor, Ontario, Canada, September 24-25 1990. International Joint Commission/Great Lakes Fishery Commission, Windsor, Ontario, Canada.

Mac, M.J., T.R. Schwartz, C.C. Edsall, and A.M. Frank. 1993. Polychlorinated biphenyls in Great Lakes trout and their eggs: Relations to survival and congener composition 1979-1988. Journal of Great Lakes Research 19:752-765.

Mace, C.P. 1997. Developing and sustaining world fisheries resources: The state of the science and management. Pp. 1-21 in D.A. Hancock, D.C. Smith, A. Grant, and J.P. Beumer, (eds.). Developing and Sustaining World Fisheries Resources. Proceedings of the 2nd World Fisheries Congress. Commonwealth Scientific Investigation and Research Organisation, Collingwood, Victoria, Australia.

Mace, C.P., and M.P. Sissenwine. 1993. How much spawning per recruit is enough? Pp. 101-118 in S.J. Smith, J.J. Hunt, and D. Rivard (eds.). Risk Evaluation and Biological Reference Points for Fisheries Management. Canadian Special Publications in Fisheries and Aquatic Sciences 120.

Mace, P.M. 1993. Will private owners practice prudent resource management? Fisheries 18(9): 29-31.

Magnien, R.E. (ed.). 1987. Monitoring for Management Actions. First Biennial Report, Water Quality Monitoring Program, Chesapeake Bay Program, Office of Environmental Programs, Maryland Department of Health and Mental Hygiene, Baltimore.

Magnusson, K.G. 1995. An overview of the multispecies VPA(theory and applications). Reviews in Fish Biology and Fisheries 5:195-212.

Man, A., R. Law, and N.V.C. Polunin. 1995. Role of marine reserves in recruitment to reef fisheries: A metapopulation model. Biological Conservation 71:197-204.

Manire, C.A., and S.H. Gruber. 1990. Many sharks may be headed toward extinction. Conservation Biology 4:10-11.

March, E.J. 1970. Sailing Trawlers. David & Charles, Newton Abbot, Devon, U.K.

Marine Mammal Commission. 1998. Annual Report To Congress 1997. Marine Mammal Commission, Bethesda, Md.

McAllister, D.E. 1996. The Status of the World Ocean and Its Biodiversity. Ocean Voice International, Ottawa.

McAllister, M.K., and R.M. Peterman. 1992. Experimental design in the management of fisheries: A review. North American Journal of Fisheries Management 12:1-18.

McCay, B.J. 1995a. Social and ecological implications of ITQs: An overview. Coastal Ocean Management 28(1-3):3-22.

McCay, B.J. 1995b. Common and private concerns. Advances in Human Ecology 4:89-116.

McCay, B.J., C.F. Creed, A.C. Finlayson, R. Apostle, and K. Mikalsen. 1995. Individual transferable quotas (ITQs) in Canadian and US fisheries. Ocean and Coastal Management 28(1-3):85-115.

McCay, B.J., and S. Jentoft. 1996. From the bottom up: Participatory issues in fisheries management. Society and Natural Resources 9(3):237-250.

McCay, B.J., and S. Jentoft. 1998. Market or community failure? Critical perspectives on common property research. Human Organization 57(1): 21-29.

McClanahan, T.R. 1994. Kenyan coral reef lagoon fish: Effects of fishing, substrate complexity, and sea urchins. Coral Reefs 13:231-241.

McClanahan, T.R., and S.H. Shafir. 1990. Causes and consequences of sea urchin abundance and diversity in Kenyan coral reef lagoons. Oecologia 362-370.

McGoodwin, J.R. 1990. Crisis in the World's Fisheries. Stanford University Press, Stanford, Calif.

Meffe, G.K., and C.R. Carroll. 1994. Principles of Conservation Biology. Sinauer Associates, Reading, Mass.

Megrey, B.A., A.B. Hollowed, S.R. Hare, S.A. Macklin, and P.J. Stabeno. 1996. Contributions of FOCI research to forecasts of year-class strength of walleye pollock in Shelikof Strait, Alaska. Fisheries Oceanography 5 (Supplement 1):189-203.

Merrett, N.R. and R.L. Haedrich. 1997. Deep-Sea Demersal Fish and Fisheries. Chapman and Hall, London.

Merritt, M.F., and K.R. Criddle. 1993. Decisions in Recreational Fisheries: A Case Study of the Chinook Salmon in the Kenai River, Alaska. Pp. 683-703 in G. Kruse, D.M. Eggers, R.J. Marasco, C. Pautzke, and T.J. Quinn II (eds.). Management Strategies for Exploited Fish Populations. Report AK-SG-93-02, Alaska Sea Grant College Program, Fairbanks.

Milazzo, M.J. 1997. Reexamining subsidies in world fisheries. In the Pacific Economic Cooperation Council Task Force on Fisheries Development and Cooperation Symposium on the Relationship Between Fisheries Management Practices and International Trade, Wellington, New Zealand, November 1996. Report of Proceedings, Ministry of Fisheries, Wellington, New Zealand.

Miles, E.L. 1994. Towards more effective management of high seas fisheries. Annual Yearbook of International Law, Vol 3:111-127.

Miller, D.M., and I. Hampton. 1989. Biology and ecology of the Antarctic krill: A review. BIOMASS Scientific Series Vol. 9. Scientific Committee on Antarctic Research, Washington, D.C.

Miller, T.J., E.D. Houde, and E.J. Watkins. 1996. Perspectives on Chesapeake Bay. Chesapeake Bay Fisheries: Prospects for Multispecies Management and Sustainability. Chesapeake Bay Program Scientific and Technical Advisory Committee. Chesapeake Research Consortium, CRC Publication No. 154B. Annapolis, Md.

Mills, E.L., J.H. Leach, J.T. Carlton, and C.L. Secor. 1993. Exotic species in the Great Lakes: A history of biotic crises and anthropogenic introductions. Journal of Great Lakes Research 19:1-54.

Moyle, P.B. 1976. Inland Fishes of California. University of California Press, Berkeley.

Muck, P. 1989. Major trends in the pelagic system off Peru and their implications for management. Pp. 386-403 in D. Pauly, M. Muck, I. Tsukayama, and J. Mendo (eds.). The Peruvian Upwelling System: Dynamics and Interactions. ICLARM Conference Proceedings, ICLARM, Manila, Philippines.

Munro, G.R. 1980. A Promise of Abundance: Extended Fisheries Jurisdiction and the Newfoundland Economy. Economic Council of Canada, Ottawa.

Munro, G.R., and A.D. Scott. 1985. The economics of fisheries management. Pp. 623-676 in A.V. Kneese and J.L. Sweeney (eds.). Handbook of Natural Resource and Energy Economics, vol. II. North-Holland, Amsterdam.

Murawski, S.A. 1991. Can we manage our multispecies fisheries? Fisheries 16(5):5-13.

Murphy, G.I. 1965. A solution of the catch equation. Journal of the Fisheries Research Board of Canada 22:191-202.

Myers, R.A., N.J. Barrowman, J.A. Hutchings, and A.A. Rosenberg. 1995. Population dynamics of exploited fish stocks at low population levels. Science 269:1106-1108.

Myers, R.A., J.A. Hutchings, and N.J. Barrowman. 1997. Why do fish stocks collapse? Ecological Applications 7(1):91-106.

Myers, R.A., and G. Mertz. 1998. The limits of exploitation: A precautionary approach. Ecological Applications 8(1) Supplement: S165-S169.

Mysak, L.A. 1986. El Niño, interannual variability and fisheries in the northeast Pacific Ocean. Canadian Journal of Fisheries Aquatic Science 43:464-497.

National Marine Fisheries Service (NMFS). 1993. World Fishing Fleets: An Analysis of Distant-Water Fleet Operations—Past-Present-Future. Vol. 1, Executive Summary, NOAA Technical Memorandum. NMFS-F/SPO-9. NMFS, Silver Spring, Md.

National Marine Fisheries Service (NMFS). 1995a. Fisheries Statistics of the United States, 1994. Current Fishery Statistics No. 9400. NMFS, U.S. Department of Commerce, Washington, D.C.

National Marine Fisheries Service (NMFS). 1995b. 1995 Individual Fishing Quota (IFQ) Allocations and Landings. National Marine Fisheries Service, Restricted Access Management Division, Juneau, AK.

National Marine Fisheries Service (NMFS). 1996a. Our Living Oceans: Report on the Status of U.S. Living Marine Resources 1995. NOAA Technical Memorandum NMFS-F/SPO-19. NMFS, Silver Spring, Md.

National Marine Fisheries Service (NMFS). 1996b. Our Living Oceans: The Economic Status of U.S. Fisheries 1996. NOAA Technical Memorandum NMFS-F/SPO-22. NMFS, Silver Spring, Md.

National Marine Fisheries Service (NMFS). 1998. Fisheries of the United States, 1997. Current Fishery Statistics No. 9700. NMFS, Silver Spring, Md.

National Marine Fisheries Service (NMFS). In press. Ecosystem-Based Fishery Management. NOAA Technical Memorandum NMFS-F/SPO-33. NMFS, Silver Spring, Md.

National Oceanic and Atmospheric Administration (NOAA). 1990. The potential of marine fishery reserves to reef fish management in the U.S. southern Atlantic. NOAA Technical Memorandum NMFS-SEFC-261. NOAA, Washington, D.C.

National Research Council (NRC). 1989. Improving Risk Communication. National Academy Press, Washington, D.C.

National Research Council (NRC). 1990. Decline of the Sea Turtles: Causes and Prevention. National Academy Press, Washington, D.C.

National Research Council (NRC). 1992a. Dolphins and the Tuna Industry. National Academy Press, Washington, D.C.

National Research Council (NRC). 1992b. Marine Aquaculture: Opportunities for Growth. National Academy Press, Washington, D.C.

National Research Council (NRC). 1994a. Environmental Information for Outer Continental Shelf Oil and Gas Decisions in Alaska. National Academy Press, Washington, D.C.

National Research Council (NRC). 1994b. An Assessment of Atlantic Bluefin Tuna. National Academy Press, Washington, D.C.

National Research Council (NRC). 1994c. Improving the Management of U.S. Marine Fisheries. National Academy Press, Washington, D.C.

National Research Council (NRC). 1994d. Priorities for Coastal Ecosystem Science. National Academy Press, Washington, D.C.

National Research Council (NRC). 1994e. Review of the Planning for the Global Ocean Observing System. National Academy Press, Washington, D.C.

National Research Council (NRC). 1994f. Science and Judgment in Risk Assessment. National Academy Press, Washington, D.C.

National Research Council (NRC). 1995a. Improving Interactions Between Coastal Science and Policy: Proceedings of the Gulf of Maine Symposium. National Academy Press, Washington, D.C.

National Research Council (NRC). 1995b. Science, Policy, and the Coast: Improving Decisionmaking. National Academy Press, Washington, D.C.

National Research Council (NRC). 1995c. Understanding Marine Biodiversity. National Academy Press, Washington, D.C.

National Research Council (NRC). 1996a. The Bering Sea Ecosystem. National Academy Press, Washington, D.C.

National Research Council (NRC). 1996b. Upstream: Salmon and Society in the Pacific Northwest. National Academy Press, Washington, D.C.

National Research Council (NRC). 1996c. Understanding Risk: Informing Decisions in a Democratic Society. National Academy Press, Washington, D.C.

National Research Council (NRC). 1996d. Stemming the Tide: Controlling Introductions of Nonindigenous Species by Ships' Ballast Water. National Academy Press, Washington, D.C.

National Research Council (NRC). 1998a. Improving Fish Stock Assessments. National Academy Press, Washington, D.C.

National Research Council (NRC). 1998b. Review of Northeast Fishery Stock Assessments. National Academy Press, Washington, D.C.

Natural Resources Consultants, Inc. 1998. Update of Discarding Practices and Unobserved Fishing Mortality in Marine Fisheries. Final report to National Marine Fisheries Service. Natural Resources Consultants, Inc., Seattle.

Nehring, D. 1991. Inorganic phosphorus and nitrogen compounds as driving forces for eutrophication in semi-enclosed areas. ICES Variability Symposium No. 44. ICES, Copenhagen.

Neis, B. 1992. Fishers' ecological knowledge and stock assessment in Newfoundland. Newfoundland Studies 8(2):155-178.

Newell, R. 1988. Ecological changes in the Chesapeake Bay: Are they the result of overharvesting the American oyster, *Crassostrea virginica*? Pp. 536-546 in M.P. Lynch and E. C. Krome (eds.). Understanding the Estuary: Advances in Chesapeake Bay Research. Proceedings of a conference, March 29-31, 1988, Baltimore. Chesapeake Research Consortium Publication No. 129. Annapolis, Md.

Nichols, F.H. 1979. Natural and anthropogenic influences on benthic community structure in San Francisco Bay. Pp. 409-426 in T.J. Conomos (ed.). San Francisco Bay: The Urbanized Estuary. American Association for the Advancement of Science, Pacific Division, San Francisco.

Nichols, F.H., J.E. Cloern, S.N. Luoma, and D. H. Peterson. 1986. The modification of an estuary. Science 231:567-573.

Nichols, F.H., J.K. Thompson, and L.E. Schemel. 1990. Remarkable invasion of San Francisco Bay (California, USA) by the Asian clam *Potamocorbula amurensis*. II. Displacement of a former community. Marine Ecology Progress Series 66:95-101.

Nixon, S.W. 1988. Physical energy inputs and the comparative ecology of lake and marine ecosystems. Limnology and Oceanography 33:1005-1025.

Norgaard, R. 1994. Development Betrayed: The End of Progress and a Coevolutionary Revisioning of the Future. Routledge, New York.

North Pacific Fishery Management Council (NPFMC). 1992. True North. NPFMC, Anchorage.

North Pacific Fishery Management Council (NPFMC). 1993. Plan Team for Groundfish Fisheries of the Bering Sea and Aleutian Islands (compilers). Stock Assessment and Fishery Evaluation Report for the Groundfish Resources of the Bering Sea/Aleutian Islands Regions as Projected for 1994. Unpublished report, NPFMC, Anchorage.

North Pacific Fishery Management Council (NPFMC). 1997. Plan Team for the Groundfish Fisheries of the Gulf of Alaska (compilers). Stock Assessment and Fishery Evaluation Report for Groundfish Resources of the Gulf of Alaska. North Pacific Fishery Management Council, Anchorage, Alaska.

Norton, Byran G. 1995. Evaluation and ecosystem management: New directions needed? The Ag Bioethics Forum 7(2):2-4.

Odendaal, F.J., M.O. Bergh, and G.M. Branch. 1994. Socio-economic options for the management of the exploitation of intertidal and subtidal resources. Pp. 155-167 in W.R. Siegfried (ed.). Rocky Shores: Exploitation in Chile and South Africa. Springer-Verlag, Heidelberg.

OECD (Organisation for Economic Co-operation and Development). 1997. Towards Sustainable Fisheries: Economic Aspects of the Management of Living Marine Resources. OECD Publications, Paris, France.

Olver, C.H., B.J. Shutter, and C.K. Minnis. 1995. Toward a definition of conservation principles for fisheries management. Canadian Journal of Fisheries Aquatic Science 52:1584-1594.

Ostrom, E. 1998. A behavioral approach to the rational choice theory of collective action. American Political Science Review 92(1):1-22.

Ostrom, E. In press. Scales, polycentricity, and incentives. In J. McNeely and L. Gurswamy (eds.). Protection of Biodiversity: Converging Interdisciplinary Strategies. Duke University Press, Durham, N.C.

Pacific Economic Council Task Force on Fisheries Development and Cooperation. 1997. Symposium on The Relationship Between Fisheries Management Practices and International Trade. Proceedings report, Pacific Economic Council, Wellington.

Paine, R.T. 1980. Food webs: Linkage, interaction strength and community infrastructure. Journal of Animal Ecology 49:667-685.

Pajaro, M.G., A.C. J. Vincent, D.Y. Buhat, and N.C. Perante. 1997. The role of seahorse fishers in conservation and management. Pp. 118-126 in Proceedings of the First International Symposium on Marine Conservation, Hong Kong.

Palumbi, S.R. 1995. Using genetics as an indirect estimator of larval dispersal. Pp. 369-387 in L. McEdward (ed.). Ecology of Marine Invertebrate Larvae. CRC Press, Boca Raton, Fla.

Parrish, R.H. 1995. Lanternfish heaven: The future of world fisheries? *Naga*, ICLARM Quarterly, July.

Parsons, T.R. 1992. The removal of marine predators by fisheries and the impacts of trophic structure. Marine Pollution Bulletin 25:51-53.

Pauly, D. 1995. Anecdotes and the shifting baseline syndrome of fisheries. Trends in Ecology and Evolution 10(10):430.

Pauly, D. 1996. One hundred million tonnes of fish, and fisheries research. Fisheries Research 25:23-38.

Pauly, D., and V. Christensen. 1995. Primary production required to sustain global fisheries. Nature 374:255-257.

Pauly, D., and T.E. Chua. 1987. The overfishing of marine resources: Socioeconomic background in Southeast Asia. Ambio 17(3):200-206.

Pauly, D., V. Christensen, J. Dalsgaard, R. Froese, and F. Torres, Jr. 1998. Fishing down marine food webs. Science 279:860-863.

Pearcy, W.G. 1992. Ocean Ecology of North Pacific Salmonids. Washington Sea Grant College Program, University of Washington, Seattle.

Pearse, P., and C. Walters. 1992. Harvesting regulation under quota management systems for ocean fisheries: Decision making in the face of natural variability, weak information, risks and conflicting incentives. Marine Policy 16:167-182.

Pennoyer, S. 1997. Bycatch management in Alaska groundfish fisheries. Pp. 141-150 in E. Pikitch, D.D. Huppert, and M. Sissenwine (eds.). Global Trends in Fisheries Management. American Fisheries Society Symposium 20. American Fisheries Society, Bethesda, Md.

Pereira, W.E., F.D. Hostettler, J.R. Cashman, and R.S. Nishioka. 1994. Occurrence and distribution of organochlorine compounds in sediments and livers of striped bass (*Morone saxatilis*) from the San Francisco Bay-Delta Estuary. Marine Pollution Bulletin 28:434-441.

Pereyra, W.T. 1996. Midwater trawls and the Alaskan pollock fishery: A management perspective. Pp. 91-94 in Alaska Sea Grant 1996. Solving Bycatch: Considerations for Today and Tomorrow. Alaska Sea Grant College Program Report 96-03, University of Alaska, Fairbanks.

Pesticide and Toxic Chemical News. 1997. *Pfiesteria piscicida* implicated in illnesses of Maryland fishermen. Volume 25 (43):8-10, August 10, 1997.

Peterson, D., D. Cayan, J. Dileo, M. Noble, and M. Dettinger. 1995. The role of climate in estuarine variability. New Scientist 83:58-67.

Pickett, S.T.A., and R. S. Ostfield. 1995. The shifting paradigm in ecology. Pp. 261-278 in R.L. Knight and S.F. Bates (eds.). A New Century for Resource Management. Island Press, Washington, D.C.

Pikitch, E. 1988. Objectives for biologically and technically interrelated fisheries. Pp. 107-136 in W.S. Wooster (ed.). Fishery Science and Management: Objectives and Limitations. Springer-Verlag, Berlin.

Pinkerton, E.W. 1994. Local fisheries co-management: A review of international experiences and their implications for salmon management in British Columbia. Canadian Journal of Fisheries and Aquatic Sciences 51:1-17.

Pinkerton, E.W. 1997. Anthropological versus economic models for control of fishing effort: Where do communities of place and communities of interest intersect? The Common Property Resource Digest 42(July):12-14.

Pinkerton, E.W., and M. Weinstein. 1995. Fisheries That Work: Sustainability Through Community-Based Management. The David Suzuki Foundation, Vancouver, B.C.

Policansky, D. 1986. North Pacific halibut fishery management. Pp.137-150 in National Research Council 1986. Ecological Knowledge and Environmental Problem-Solving: Concepts and Case Studies. National Academy Press, Washington, D.C.

Policansky, D. 1993a. Fishing as a cause of evolution in fishes. Pp. 2-18 in T.K. Stokes, J.M. McGlade, and R. Law (eds.). The Exploitation of Evolving Resources. Lecture Notes in Biomathematics, vol. 99. Springer-Verlag, Berlin.

Policansky, D. 1993b. Evolution and management of exploited fish populations. Pp. 650-664 in G. Kruse, D.M. Eggers, R.J. Marasco, C. Pautzke, and T.J. Quinn II (eds.). Management Strategies for Exploited Fish Populations. Report AK-SG-93-02, Alaska Sea Grant College Program, Fairbanks.

Policansky, D., and J.J. Magnuson. 1998. Genetics, metapopulations, and ecosystem management of fisheries. Ecological Applications 8(1) Supplement: S119-S123.

Polovina, J.J., and W. R. Haight. In press. Climate variation, ecosystem dynamics, and fisheries management in the northwestern Hawaiian Islands. In P. Muller and G. Holloway, eds. Biotic Impacts of Extratropical Climate Variability in the Pacific. Proceeding 'Aha Huliko'a Hawaiian Winter Workshop, University of Hawaii at Manoa, January 1998.

Polunin, N.V.C., and C.M. Roberts. 1993. Greater biomass and value of target coral-reef fishes in two small Caribbean marine reserves. Marine Ecology Progress Series 100:167-176.

Porter, G. 1997. Fishing subsidies, overfishing, and trade. Background paper prepared for the UNEP/WWF Workshop on the Role of Trade Policies in the Fishing Sector, Geneva, June 2-3, 1997. World Wildlife Fund, Washington, D.C.

Reingold, H. 1993. The Virtual Community: Homesteading on the Electronic Frontier. Addison Wesley, Reading, Mass.

Richkus, W.A., H.M. Austin, and S.J. Nelson. 1992. Fisheries assessment and management synthesis: Lessons for Chesapeake Bay. Chesapeake Bay Program, Scientific and Technical Advisory Committee. Perspectives on Chesapeake Bay, 1992: Advances in Estuarine Science 75-114. Chesapeake Research Consortium Publication No. 143. Annapolis, Md.

Ricker, W.E. 1954. Stock and recruitment. Journal of the Fisheries Research Board of Canada 11:559-623.

Ricker, W.E. 1958. Maximum sustained yields from fluctuating environments and mixed stocks. Fisheries Research Board of Canada Bulletin 119, Fisheries Research Board of Canada, Ottawa.

Ricker, W.E. 1975. Computation and Interpretation of Biological Statistics of Fish Populations. Fisheries Research Board of Canada Bull. 191, Fisheries Research Board of Canada, Ottawa.

Rieser, A. 1991. Ecological preservation as a public property right: An emerging doctrine in search of a theory. Harvard Environmental Law Review 15:273-313.

Rieser, A. 1997. Property rights and ecosystem management in U.S. fisheries: Contracting for the commons? Ecology Law Quarterly 24(4):813-832.

Rijnsdorp, A. 1992. Long-Term Effects of Fishing in North Sea Plaice: Disentangling Genetic and Phenotypic Plasticity in Growth, Maturation, and Fecundity. Ph.D. thesis, University of Amsterdam, Netherlands.

Robert, C., W.J. Ballantine, C.D. Buxton, L.B. Crowder, W. Milon, M.K. Orbach, D. Pauly, J. Trexler, and C.J. Walters. 1995. Review of the use of marine fishery reserves in the U.S. Southeastern Atlantic NOAA Technical Memorandum. NMFS-SEFCS-376, NMFS, Silver Spring, Md.

Roberts, C. M. 1997. Ecological advice for the global fisheries crisis. Trends in Ecology and Evolution 12:35-38.

Roberts, C.M., and N.V.C. Polunin. 1991. Are marine reserves effective in management of reef fisheries? Reviews in Fish Biology and Fisheries 1:65-91.

Roberts, C.M., and N.V.C. Polunin. 1993a. Marine reserves: Simple solutions to managing complex fisheries? Ambio 22:363-368.

Roberts, C.M., and N.V.C. Polunin. 1993b. Effects of marine reservation protection on northern Red Sea fish populations. Proc 7th International Coral Reef Symposium. Smithsonian Tropical Research Institution, Balboa, Panama.

Rosenberg, A., M. Fogarty, M. Sissenwine, J. Beddington, and J. Shepherd. 1993. Achieving sustainable use of renewable resources. Science 262:828-829.

Rothschild, B. 1983. On the allocation of fisheries stocks. Pp. 85-91 in J.W. Reintkes (ed.). Improving Multiple Use of Coastal and Marine Resources. American Fisheries Society, Bethesda, Md.

Rothschild, B.J. 1986. Dynamics of Marine Fish Populations. Harvard University Press, Cambridge, Mass.

Rothschild, B. 1995. Fishstock fluctuations as indicators of multidecadal fluctuations in the biological productivity of the ocean. Pp. 201-209 in R.J. Beamish (ed.). Climate Change and Northern Fish Populations. Canadian Special Publications in Fisheries and Aquatic Sciences 121.

Rothschild, B.J., J.S. Ault, P. Goulletquer, and M. Heral. 1994. Decline of the Chesapeake Bay oyster population: A century of habitat destruction and overfishing. Marine Ecology Progress Series 111:29-39.

Rowley, R.J. 1994. Case studies and reviews: Marine reserves in fisheries management. Aquatic Conservation: Marine and Freshwater Ecosystems 4:233-254.

Ruckelshaus, M., and C. Hays. 1997. Conservation and management of species in the sea. Pp. 112-156 in P.L. Fiedler and P.M. Kareiva, eds. Conservation Biology for the Coming Decade, 2nd. ed. Chapman and Hall, New York.

Russ, G.R. 1985. Effect of protective management on coral reef fishes in the central Philippines. Proceedings of the 5th International Coral Reef Congress. 4:219-224.

Russ, G.R. 1989. Distribution and abundance of coral reef fishes in the Sumilon Island Reserve, central Philippines, after nine years of protection from fishing. Asian Marine Biology 6:59-71.

Russ, G.R., and A.C. Alcala. 1989. Effects of intense fishing pressure on an assemblage of coral reef fisheries. Marine Ecology Progress Series 56:13-27.

Russ, G.R., and A.C. Alcala. 1994. Sumilon Island Reserve: 20 years of hopes and frustrations. *Naga*, the ICLARM Quarterly 7(3):8-12.

Russ, G.R., and A.C. Alcala. 1996. Marine reserves: rates and patterns of recovery and decline in abundance of large predatory fish. Ecological Applications 6(3):947-961.

Ryther, J.H. 1969. Photosynthesis and fish production in the sea. Science 166:72-76.

Safina, C. 1995. The world's imperiled fish. Scientific American 273(5):46-53.

Samoilys, M. 1988. Abundance and species richness of coral reef fish on the Kenyan coast: The effects of protective management and fishing. Proceedings of the 6th International Coral Reef Symposium 2:261-266.

Sand, P.H. (ed.). 1992. The Effectiveness of International Environmental Agreements: A Survey of Existing Legal Instruments. Grotius Publications Ltd., Cambridge, England.

Santelices, B. 1989. Algas Marinas de Chile: Distribucion, Ecologia, Utilizacion, Diversidad. Ediciones Universidad Catolica de Chile, Santiago, Chile.

Schaefer, M.B. 1965. The potential harvest of the sea. Transactions of the American Fisheries Society 94:123-128.

Schreiner, D.R., and S.T. Schram. 1997. Lake trout rehabilitation in Lake Superior. Fisheries 22 (7):12-14.

Scofield, N.B., and H.C. Bryant. 1926. The striped bass in California. California Fish Game Bulletin 12(2):55-74.

Scott, A. 1993. Obstacles to fishery self-government. Marine Resource Economics 8:187-199.

Sharp, G.D., and D.R. McLain. 1995. Fisheries, El Niño-Southern Oscillation and upper temperature records: An eastern Pacific example. Oceanography 6:13-22.

Sherman, K. 1990. Productivity, perturbations, and options for biomass yields in large marine ecosystems. Pp. 206-219 in K. Sherman, L.M. Alexander and B.D. Gold (eds.), Large Marine Ecosystems: Patterns, Processes and Yields. AAAS Press, Washington, D.C.

Sherman, K., L.M. Alexander, and B.D. Gold (eds.). 1993. Large Marine Ecosystems: Stress, Mitigation and Sustainability. AAAS Press, Washington, D.C.

Siegel, V., and V. Loeb. 1995. Recruitment of Antarctic krill, Euphausia superba, and possible causes for its variability. Marine Ecology Progress Series 123:45-56.

Simenstad, C.A., J.A. Estes, and K.W. Kenyon. 1978. Aleuts, sea otters, and alternate stable-state communities. Science 200:403-411.

Sissenwine, M.P. 1978. Is MSY an adequate foundation for optimum yield? Fisheries 3(6):22-42.

Sissenwine, M.P. 1984. Why do fish populations vary? Pp. 59-94 in R.M. May (ed.). Exploitation of Marine Communities. Dahlem Konferenzen, Springer-Verlag, Berlin.

Sissenwine, M.P., and N. Daan. 1991. An overview of multispecies models relevant to the management of living marine resources. International Council for Exploration of the Sea Marine Science Symposia 193:6-11

Sissenwine, M.P., and P.M. Mace. 1992. ITQs in New Zealand: The era of fixed quotas in perpetuity. Fishery Bulletin, U.S. 90:147-160.

Sissenwine, M.P., and A. Rosenberg. 1993. Marine fisheries at a critical juncture. Fisheries 18(10):6-14.

Sissenwine, M.P., E.B. Cohen, and M.D. Grosslein. 1984. Structure of the Georges Bank ecosystem. Rapport Proces-Verbaux Reunion, Conseil Internationale Explorations de la Mer 183:243-254.

Skinner, J.E. 1962. An Historical Review of the Fish and Wildlife Resources of the San Francisco Bay Area. California Fish and Game Water Projects Branch Report 1, California Fish and Game, Sacramento.

Skjoldal, H.R., H. Gjter, and H. Loeng. 1992. The Barents Sea ecosystem in the 1980s: Ocean climate, plankton, and capelin growth. International Council for Exploration of the Sea Marine Science Symposia 195:278-290.

Smith, S. 1997. Giving IFQs their due. Editorial. The National Fisherman, June, p. 7.

Smith, T. 1994. Scaling Fisheries. Cambridge University Press, Cambridge, England.

Soutar, A., and J.D. Isaacs. 1974. Abundance of pelagic fish during the 19th and 20th centuries as recorded in anaerobic sediments off California. Fish Bulletin of the U.S. Fish and Wildlife Service 72:257-275.

Sparre, P. 1991. Introduction to multispecies virtual population analysis. International Council for Exploration of the Sea Marine Science Symposia 193:12-21.

Spratt, J.D. 1981. The Status of the Pacific Herring, *Clupea harengus pallasii*, Resource in California 1972 to 1980. California Department of Fish and Game, Fishery Bulletin 171.

Spratt, J.D. 1992. Pacific Herring. Pp. 86-89 in W.S. Leet, C.M. Dewees, and C.D. Hougan (eds.). California's Living Marine Resources and their Utilization. California Sea Grant Extension Publication UCSGEP-92-12, Davis, Calif.

Steele, D.H., R. Andersen, and J.M. Green. 1992. The managed commercial annihilation of the northern cod. Newfoundland Studies 8(1):34-68.

Steele, J.H. 1985. A comparison of terrestrial and marine ecological systems. Nature 313:355-358.

Steele, J.H. 1991a. Marine ecosystem dynamics: comparison of scales. Ecological Research 6:175-183.

Steele, J.H. 1991b. Marine functional diversity. Bioscience 41:470-474.

Steele, J.H. 1991c. Can ecological theory cross the land-sea boundary? Journal of Theoretical Biology 153:425-436.

Steele, J.H. 1996. Regime shifts in fisheries management. Fisheries Research 25:19-23.

Steele, J.H. 1998. Regime shifts in marine ecosystems. Ecological Applications 8(1) Supplement: S33-S36.

Sutherland, J.P. 1974. Multiple stable points in natural communities. American Naturalist 108:859-873.

Swezey, S.L., and R.F. Heizer. 1977. Ritual management of salmonid fish resources in California. The Journal of California Anthropology 4(1):6-29.

Task Force on Atlantic Fisheries (the "Kirby report"). 1983. Navigating Troubled Waters: A New Policy for the Atlantic Fisheries. Ministry of Supply and Services, Ottawa.

Tegner, M.J., and P.K. Dayton. 1981. Population structure, mortality, and recruitment of two sea urchins (*Strongylocentrotus franciscanus* and *S. purpuratus*) in a kelp forest. Marine Ecology Progress Series 5:255-268.

Thompson, G. 1993. Compensating for harvest externalities in the management of interjurisdictional fisheries. Pp. 721-743 in G. Kruse, D.M. Eggers, R.J. Marasco, C. Pautzke, and T.J. Quinn II (eds.). Management Strategies for Exploited Fish Populations. Report AK-SG-93-02, Alaska Sea Grant College Program, Fairbanks.

Thompson, W.F. 1919. The scientific investigation of marine fisheries, as related to the work of the Fish and Game Commission in Southern California. Fisheries Bulletin (California) 2:3-27.

Thompson, W.F., and N.L. Freeman. 1930. History of the Pacific Halibut Fishery. Report of the International Fisheries Commission (International Pacific Halibut Commission). Wrigley Printing Co., Vancouver, B.C.

Townsend, R.E. 1995. Fisheries self-governance: Corporate or cooperative structures? Marine Policy 19(1).

Townsend, R.E., and S.G. Pooley. 1995. Corporate management and the northwestern Hawaiian islands. Ocean and Coastal Management 28(1):63-83.

Trumble, R.J. 1996. Management of Alaskan longline fisheries to reduce halibut bycatch mortality. Pp. 183-192 in Alaska Sea Grant 1996. Solving Bycatch: Considerations for Today and Tomorrow. Alaska Sea Grant College Program Report 96-03, University of Alaska, Fairbanks.

Trumble, R.J., and R.D. Humphries. 1985. Management of Pacific herring (*Clupea harengus pallasi*) in the eastern Pacific Ocean. Canadian Journal of Fisheries and Aquatic Sciences 42(Supplement 1):230-244.

Tseng, C.K., ed. 1984. Common Seaweeds of China. Science Press, Beijing; Kruger Publications, Amsterdam and Berkeley, Calif.

Ulanowicz, R.E., and J.H. Tuttle. 1992. The trophic consequences of oyster stock rehabilitation in Chesapeake Bay. Estuaries 15(3):298-306.

Ulltang, O. 1980. Factors affecting the reaction of pelagic fish stocks to exploitation and requiring a new approach to assessment and management. Rapports Proces-Verbaux Reunion, Conseil Internationale Exploration de la Mer 177:489-504.

United States Census Bureau. 1998. International Data Base Web Site http://www.census.gov/ipc/www/img/worldpop.gif.

USGS (United States Geological Survey). 1998. Web site http://sfbay.wr.usgs.gov/access/bruce/intro.html.

Van Blaricom, G.R., and J.A. Estes (eds.). 1988. The Community Ecology of Sea Otters. Springer-Verlag, Berlin.

Van der Elst, R.P. 1979. A proliferation of small sharks in the shore-based Natal sports fishery. Environmental Biology of Fishes 4:349-362.

Vincent, A.C.J. 1996. The International Trade in Seahorses. A TRAFFIC Network Report. TRAFFIC International, Cambridge, England.

Vincent, A.C.J. 1997a. Implementing marine reserves with subsistence communities. Keynote paper for the University of British Columbia Fisheries Centre Workshop on the Design and Monitoring of Marine Reserves, University of British Columbia, Vancouver, B.C., Canada. Abstract published in The Design and Monitoring of Marine Reserves, University of British Columbia Fisheries Centre Research Reports, 1997, vol. 5 no. 1. University of British Columbia Fisheries Centre, Vancouver.

Vincent, A.C.J. 1997b. Sustainability of seahorse fishing. Pp. 2045-2050 in H.A. Lessios and I.G. Macintyre (eds.). Proceedings of the 8th International Coral Reef Symposium, Volume 2. Smithsonian Tropical Research Institute, Balboa, Panama.

Walters, C. 1986. Adaptive Management of Renewable Resources. MacMillan, New York.

Walters, C., V. Christensen, and D. Pauly. 1997. Structuring dynamic models of exploited ecosystems from trophic mass-balance assessments. Reviews in Fish Biology and Fisheries 7:1-34.

Walters, C., R.D. Goruk, and D. Radford. 1993. Rivers Inlet sockeye salmon: An experiment in adaptive management. North American Journal of Fisheries Management 13:253-262.

Walters, C., and R. Hilborn. 1976. Adaptive control of fishing systems. Journal of the Fisheries Research Board of Canada 33:145-159.

Walters, C., and J-J. Maguire. 1996. Lessons for stock assessment from the northern cod collapse. Reviews in Fish Biology and Fisheries 6(2):125-138.

Walters, C., and A.M. Parma. 1996. Fixed exploitation rate strategies for coping with effects of climate change. Canadian Journal of Fisheries and Aquatic Sciences 53:148-158.

Ware, D.M. 1985. Life history characteristics, reproductive value and resilience of Pacific herring (*Clupea harengus pallasi*). Canadian Journal of Fisheries and Aquatic Sciences 42 (Supplement 1):127-137.

Wilson, J.A., J. French, P. Kleban, S.R. McKay, and R. Townsend. 1991. Chaotic dynamics in a multiple species fishery: A model of community precision. Ecological Modeling 58:303-322.

Wooster, W.S. (ed.). 1983. From Year to Year: Interannual Variability of the Environment and Fisheries of the Gulf of Alaska and the Eastern Bering Sea. Washington Sea Grant College Program, University of Washington Press, Seattle.

Wooster, W.S., and D.L. Fluharty (eds.). 1985. El Niño North: Niño Effects in the Eastern Subarctic Pacific Ocean. Washington Sea Grant College Program, University of Washington Press, Seattle.

World Wildlife Fund (WWF). 1995. Managing U.S. Marine Fisheries: Public Interest or Conflict of Interest? World Wildlife Fund, Washington, D.C.

World Wildlife Fund (WWF). 1997. Subsidies and Depletion of World Fisheries: Case Studies. World Wildlife Fund, Washington, D.C.

Yablokov, A.V. 1994. Validity of whaling data. Nature 367:108.

Yamanaka, I. 1983. Interaction among krill, whales, and other animals in the Antarctic ecosystem. Memoirs of the National Institute of Polar Research. 27 (Special Issue):220-232.

Yamasaki, A., and A. Kuwhara. 1990. Preserved areas do affect recovery of overfished crab stocks of Kyoto Prefecture. Pp. 575-578 in Proceedings of the International Symposium on King and Tanner Crabs. Alaska Sea Grant, University of Alaska, Fairbanks.

Young, M.D., and B.J. McCay. 1995. Building equity, stewardship, and resilience into market-based property right systems. Pp. 87-102 in S. Hanna and M. Munasinghe (eds.). Property Rights and the Environment: Social and Ecological issues. The World Bank, Washington D.C.

A
Papers Presented at International Conference on Ecosystem Management for Sustainable Marine Fisheries, Monterey, California February 19-24, 1996[1]

Arnason, Ragnar, University of Iceland. Ecological fisheries management using individual transferable share quotas.

Bingham, Nathaniel, Pacific Coast Federation of Fishermen's Associations. Humans impact both resources and the environment.

Burkholder, JoAnn, North Carolina State University. How do blooms of harmful microalgae and heterotropic dinoflagellates impact management of sustainable fisheries?

Castilla, Juan Carlos, and Miriam Fernandez, Pontificia Universidad de Chile. Small-scale benthic fisheries in Chile: a lesson on co-management and sustainable use of benthic invertebrates.

Christensen, Villy, and Daniel Pauly, University of British Columbia. Marine ecosystem management: an ode to Odum.

Clark, Colin, University of British Columbia. Refugia.

Clark, Jerry, National Fish and Wildlife Foundation, and Dayton Lee Alverson, Natural Resources Consultants, Inc. Are present resource management systems adequate?

Clark, Jerry, National Fish and Wildlife Foundation. Case study: Atlantic salmon.

Clark, Jerry, National Fish and Wildlife Foundation. Case study: Shrimp/ red snapper.

Dewees, Christopher, University of California. Summary of individual quota systems and their effects on New Zealand and British Columbia fisheries.

---

[1]Many of these papers have been published in Ecological Applications 8(1), Supplement (1998).

Done, Terry, and R.E. Reichelt, Australian Institute of Marine Science. The role of integrated coastal zone management in achieving sustainable marine fisheries through managing ecosystems.

Folke, Carl, and Nils Kautsky, Stockholm University. The ecological footprint concept for sustainable seafood production.

Fujita, Rodney, and Ianthe Zevos, Environmental Defense Fund. Innovative approaches for fostering conservation in marine fisheries.

Fujita, Rodney, Environmental Defense Fund. Case study: Restoring the ecological integrity of the South Florida landscape/seascape.

Hofmann, Eileen, Old Dominion University, and Thomas Powell, University of California. The environment affects biological systems.

Hofmann, Eileen, Old Dominion University. Case study: Antarctic krill fishery.

Houde, Edward, Chesapeake Biological Laboratory, and Ellen Pikitch, University of Washington. Biological resources interact in complex ways.

Houde, Edward, Chesapeake Biological Laboratory. Case study: Sustaining fisheries in the Chesapeake Bay.

Kaly, U.L., James Cook University, Australia. Mangrove restoration: a potential ecosystem management tool for sustainable development.

Kurien, John, Center for Developmental Studies, India. Old proverbs of Asian coastal communities.

Lubchenco, Jane, Oregon State University, and Simon Levin, Princeton University. The importance and challenge of sustainability.

Magnuson, John, University of Wisconsin. Case study: Laurentian Great Lakes.

Magnuson, John, University of Wisconsin. Case study: Salmon in the Pacific Northwest.

Matsuda, Hiroyuki, et al., Kyushi University. Mathematical ecology for sustainable use of the plankton-feeding pelagic fishes.

McCay, Bonnie, Rutgers University. Humans impact both resources and the environment.

McCay, Bonnie, Rutgers University. Case study: ITQs in two fisheries.

Munro, Gordon, University of British Columbia. Humans impact both resources and the environment: an economist's perspective.

Munro, Gordon, University of British Columbia. Case study: Northern cod.

National Research Council (1996). Case study: The Bering Sea ecosystem (discussion of a report).

Paine, Robert, University of Washington. Case study: Sea otters.

Pauly, Daniel, University of British Columbia; and Michael Sissenwine, National Marine Fisheries Service, Woods Hole. The nature of marine ecosystems.

Pauly, Daniel, University of British Columbia. Case study: Developing on the edge of an abyss: the fisheries of the Philippines.

Pinkerton, Evelyn, University of British Columbia. The role of dispute resolution in new paradigms for marine ecosystem management.

Pope, John G., and J.W. Horwood, Ministry of Agriculture, Fisheries, and Food, United Kingdom. Monitoring the marine ecosystem for sustainable fisheries: East Atlantic approaches.

Roughgarden, Jonathan, Stanford University, California. How to manage fisheries.

Sissenwine, Michael, National Marine Fisheries Service, Woods Hole. Sustaining fisheries by ecosystem management.

Spratt, J.D. Case study: The herring fishery in San Francisco Bay (discussion of a paper).

Steele, John H., Woods Hole Oceanographic Institution. Marine ecosystems: sustainable for fungible?

Steele, John H., Woods Hole Oceanographic Institution. Regime shifts in fisheries management.

Williams, Meryl, ICLARM, Philippines. Case study: Shrimp and mangroves.

APPENDIX
B
Biographies of Committee and Staff

Harold Mooney (chair) is the first Paul S. Achilles Professor of Biology at Stanford University and the author of numerous publications on ecology. Dr. Mooney's research includes the impacts of global change on terrestrial ecosystems in the desert, temperate, mediterranean, and tropical zones. He received his Ph.D. from Duke University. Dr. Mooney is a member of the National Academy of Sciences and is a fellow of the American Association for the Advancement of Science.

Nathaniel Bingham was the habitat director of the Pacific Coast Federation of Fishermen's Associations. He had also worked as a fisheries consultant since 1992, including contracts with Pacific States Marine Fisheries Commission Habitat education program and the Bay Institute. Before that, Mr. Bingham owned and operated a vessel and fished for salmon, Dungeness crab, and albacore.

Dayton Lee Alverson earned his Ph.D. from the University of Washington. Dr. Alverson is President of Natural Resources Consultants and is an adjunct professor at the University of Washington. His research interests include fisheries, management, population diagnosis, and resource distribution and behavior, with a recent focus on bycatch. Dr. Alverson chaired an independent panel reviewing Canadian stock assessments of northern cod.

Jerry Clark earned his Ph.D. in natural resource economics from Oregon State University. He is Director of Fisheries for the National Fish and

Wildlife Foundation in Washington, D.C. His past employment and research have focused on economics and statistics.

Frederick Grassle earned a Ph.D. in zoology from Duke University. Dr. Grassle is Professor of Marine and Coastal Sciences and Director of the Institute of Marine and Coastal Sciences of Rutgers University. His research focuses on marine benthic organisms, populations and ecosystem processes in estuarine, coral reef, continental shelf, and deep-sea environments.

Eileen Hofmann earned her Ph.D. in Marine Sciences and Engineering at North Carolina State University. She is a Professor in the Department of Oceanography at Old Dominion University. Dr. Hofmann's research includes analysis and modeling of biological and physical interactions in marine ecosystems and descriptive physical oceanography.

Edward Houde earned his Ph.D. in fisheries science from Cornell University. He is Professor at the Chesapeake Biology Laboratory of the University of Maryland. Dr. Houde's research includes ecology and developmental biology of fish eggs and larvae, assessment of pelagic resource abundance, recruitment processes, and factors leading to fluctuations in fish stock abundances.

Simon Levin earned his Ph.D. from the University of Maryland. Dr. Levin is the George M. Moffett Professor of Biology and the Director of the Princeton Environmental Initiative at Princeton University. His research focuses on theoretical ecology; mathematical and computational models of ecological and evolutionary processes; and terrestrial, intertidal, and marine ecosystems.

Jane Lubchenco earned her Ph.D. in ecology from Harvard University. Dr. Lubchenco is Wayne and Gladys Valley Professor of Marine Biology at Oregon State University, as well as a Research Associate of the Smithsonian Institution. She is a member of the National Academy of Sciences and is past president of the American Association for the Advancement of Science. Her research focuses on population and community ecology, marine ecology, algal ecology, algal life histories, and chemical ecology.

John Magnuson serves as professor of zoology and director of the Center for Limnology at the University of Wisconsin, Madison. He earned his Ph.D. in zoology from the University of British Columbia and BSc and Msc. from the University of Minnesota. His research interests include long-term regional ecological of aquatic systems, climate-change effects on lake ecosystems, fish and fisheries ecology, community ecology of lakes as islands, and ecology of the Great Lakes.

Bonnie McCay earned her Ph.D. in anthropology from Columbia University. Dr. McCay is Professor at the Department of Human Ecology and Associate Director of the Ecopolicy Center at Cook College of Rutgers University. Her research deals with the socioeconomic, cultural, and political dimensions of marine fisheries and fisheries management as well as community responses to industrialization related to offshore oil and gas production.

Gordon Munro earned his Ph.D. in economics from Harvard University. He is Professor in the Economics Department of the University of British Columbia. His research focuses on natural resource economics, particularly for renewable resources.

Robert Paine earned a Ph.D. in zoology from the University of Michigan. He is Professor of Zoology at the University of Washington and a member of the National Academy of Sciences. Dr. Paine's research focuses on algal ecology, prey-predator relationships, and community organization.

Stephen Palumbi earned his Ph.D. from the University of Washington. He is an Associate Professor at Harvard University. Dr. Palumbi's research focuses on population biology and evolutionary ecology of marine species, including an interest in the theory and practice of marine protected areas.

Daniel Pauly earned his doctorate degree from the University of Kiel, Germany, in fisheries biology, zoology and oceanography. He is Professor at the Fisheries Centre, University of British Columbia, and is also the Principal Science Adviser for the International Centre for Living Aquatic Resource Management (ICLARM) in the Philippines. Dr. Pauly's research focuses on tropical fisheries management, ecosystem modeling, and comparative studies of growth and related processes in wild and cultivated fish.

Ellen Pikitch earned a Ph.D. in zoology from Indiana University. She is Director of Fish Conservation Programs at the Wildlife Conservation Society. Dr. Pikitch's main research interests are in fisheries science, stock assessment, bycatch problems, and other marine conservation issues.

Thomas Powell earned his Ph.D. in physics from the University of California, Berkeley. Dr. Powell is Professor in the Department of Integrative

Biology at the University of California, Berkeley. His research interests are the impact of physical processes (such as currents, waves, and mixing) on the ecology of plankton in lakes, estuaries, and the coastal ocean; and measurement and modeling of physical and biological processes.

Michael Sissenwine earned a Ph.D. in oceanography from the University of Rhode Island. Dr. Sissenwine is the director of the Northeast Fisheries Science Center. His research focuses on fish-population dynamics, fishery-management systems, trophic interrelationships in marine ecosystems, and biological systems models and simulations.

NRC Staff

David Policansky (project director) earned a Ph.D. in biology from the University of Oregon. Dr. Policansky is associate director of the National Research Council's Board on Environmental Studies and Toxicology. He has staffed studies on natural resources, ecology, and fisheries at the National Research Council and his research interests include population ecology, fisheries science, and application of science to policy.

Index